LIVESTOCK

SHOWMAN'S

HANDBOOK

LIVESTOCK SHOWMAN'S HANDBOOK

A GUIDE FOR RAISING ANIMALS FOR JUNIOR LIVESTOCK SHOWS

ROGER POND

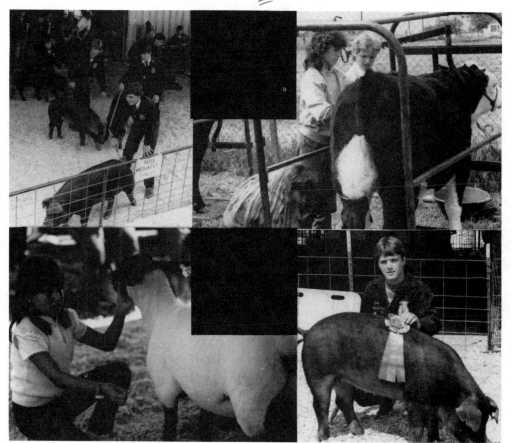

PINE FOREST PUBLISHING

Library of Congress # 86-63395

ISBN # 0-9617766-0-9

Published by Pine Forest Publishing
314 Pine Forest Road
Goldendale, Washington 98620

Printed in the United States of America

Acknowledgments

While it would be impossible to mention all of those who have contributed to this book, the author would like to recognize a few who have provided considerable information and assistance.

These include the many county agents who are the source of much of the information included in the book. Special thanks goes to Eddie Thomason for writing assistance and advice with materials about dairy cattle, and Leath Andrews for his assistance in writing and compiling swine breeding stock information. Thanks also goes to John Fouts, who has answered many questions — and spent countless hours searching the files for information the author was sure he had seen somewhere.

The author is grateful to Gwen and Theo Caldwell for their expert advice on all aspects of sheep production and for providing excellent photo opportunities.

He wishes to thank veterinarians, Jerry Harsch, Carl Conroy, and Rob Shimek, who are very generous with their knowledge and subscribe to the theory that educating clients is good business.

A special thanks is also due Tom Henry, Norm Herdrich, and staff at Western *Farmer-Stockman* magazines, where most of the book's information originally appeared in the author's *Growing & Showing* column. Information for magazine columns included in the book was also contributed by a number of livestock breeders including Neil and Jill Kayser, Sharon and Jim Pond, and others too numerous to list.

He is also grateful to Craig Steindorf for assistance in editing text and to Connie Pond for design of the book from cover to cover — and for photographing all of those animals to which she is allergic.

DISCLAIMER

Every effort has been made to make this book as complete and accurate as possible. The book is not meant to be the last word in feeding and care of animals, however.

Information in the book is derived from university recommendations and research results wherever possible; but there may be mistakes in print or in content. This book is not intended to give advice on veterinary procedures or treatments. Competent professionals in veterinary medicine, or nutrition should be consulted if advice in these fields is desired.

The author and Pine Forest Publishing shall assume neither responsibility nor liability for any losses or damages caused or said to be caused by following any information or recommendations contained in this book.

TABLE OF CONTENTS

Part III SWINE

Part IV DAIRY CATTLE

Part V FEEDS AND RATIONS

Part VI JUNIOR SHOWS

APPENDIX

Introduction

Animals are one of the few commodities subject to reverse heredity. Like insanity, one often gets them from the kids.

A bad case of animals strikes one out of two families involved in 4-H and FFA programs around the country, and although there is a wealth of good information on raising livestock, it is seldom where you can find it. The 4-H leader and the family with only a few animals needs a quick reference and background information to help understand the basics of feeding, fitting, and showing junior livestock projects.

This book attempts to meld scientific information of value to all livestock producers with the special needs of the junior livestock member, small scale producer, and backyard sheepherder. The animals have the same basic needs regardless of how many we may own, but the facilities and techniques for meeting these needs will differ.

The author's technical training and more than 15 years experience as a county extension agent and vocational agriculture teacher in Ohio and Washington provide valuable insight into the problems faced by junior livestock producers, 4-H leaders and vocational agriculture advisors. *LIVESTOCK SHOWMAN'S HANDBOOK* combines scientific livestock production information with personal experiences of the author and others in an easily understood and readable form.

Much of this information was compiled from the author's column, *Growing & Showing*, which has appeared in several magazines and farm publications around the U.S., including the *Farmer-Stockman* magazines of Washington, Oregon, Idaho, and Montana; *Today's Farmer* circulated in Missouri and surrounding states; *Family Farm* with circulation in the northeastern states; and *Farmweek* of Indiana and Ohio.

BEEF
CATTLE

Selecting Market Steer Projects

Market steer selection is the crucial first step in producing a winning entry for any steer show. This holds true for the youth show at a small county fair or the no-holds-barred open class competition at the Cow Palace in San Francisco. You can mix all sorts of special rations and feed like a pro; but if your steer isn't correct in conformation and size, you'll still have that blank space in the trophy cabinet.

I hasten to add that having the grand champion steer is not at the top of everyone's list of goals. At least I hope it isn't. I believe most 4-H and FFA members and parents value the learning experience of the project above having the one top animal.

I have always thought those who measure success in terms of grand champions are almost certainly headed for disappointment. The reason the grand champion is the grand champion is because the judge says he is. Kids generally accept this fact. It's something to strive for, but let's keep it in perspective.

I'll also toss in the observation that many 4-H and FFA members achieve satisfaction from showing animals they raised themselves or that Dad produced on the ranch. They may even prefer the cute one with the floppy ears. Aghhh!

With due respect to the uncertainties of steer shows, there are some guidelines to help the young showman select a steer and a feeding program which will provide a chance for a top placing. The first requirement is to select a steer of good conformation and correct size and age for the show you have in mind. Most youth shows have a minimum weight limit; some also have a maximum weight. We would like a steer that can fit into those limits and meet the demands of the current market.

There's always some debate about whether shows are in tune with current markets, but most show judges seem to be looking for a steer in the 1,100 to 1,300 pound range. Many judges and packers would say the 1,300 pound steer is too big for today's market, but judges have to work with what's at the show.

We should remember that the showman has to feed this animal for several months, and coming to the fair with exactly the weight you had in mind isn't easy. Since the recommended size and type of steer will vary in different parts of the country; it's a good idea to check on the show weights of winning steers in your area the last few years.

Just for example, let's say we want a steer weighing 1,150 to 1,200 pounds at a show which is six months away. What size should we be looking for?

We will begin by looking for a steer with enough frame to grow to the desired weight without being overfat and one which is likely to grade choice at this weight. Experienced cattlemen do this by virtue of their trained eye and almost certain knowledge that they won't be around after the show or that you won't remember what they said if their estimate falls a little short.

The less experienced may benefit from the use of a chart such as the Missouri Frame Score Chart, which I have included here. This chart appeared in *Successful Farming* magazine back in 1982.

You can see this chart estimates the proper slaughter weight for individual steers based upon the age and hip height of the animal at the time the measurement is taken. In other words, it predicts the weight at which a steer of a particular frame size can be expected to grade choice.

The chart is used by taking a hip height measurement in inches with the calf standing on level ground. One simply lays a yardstick across the hips of the steer and measures from that point to the ground. Putting the calf in a scales or chute to do this works well.

4

Expected Slaughter Wt.		750-850	851-950	951-1050	1051-1150	1151-1250	1251-1350	1350+
Frame Size		1	2	3	4	5	6	7
AGE IN MONTHS	7	36	38	40	42	44	46	48
	8	37	39	41	43	45	47	49
	9	38	40	42	44	46	48	50
	10	39	41	43	45	47	49	51
	11	40	42	44	46	48	50	52
	12	41	43	45	47	49	51	53
	13	41.5	43.5	45.5	47.5	49.5	51.5	53.5
	14	42	44	46	48	50	52	54
	15	42.5	44.5	46.5	48.5	50.5	52.5	54.5
	16	43	45	47	49	51	53	55
	17	43.5	45.5	47.5	49.5	51.5	53.5	55.5
	18	44	46	48	50	52	54	56

The height in inches shown under each frame size is the minimum height for that frame size.

This measurement and the calf's age is applied to the chart to determine the animal's frame size and expected slaughter weight. Looking at the chart, you can see if the calf is 12 months old and measures 43 inches at the hip, he is a frame size two and would be expected to finish at a slaughter weight of 850 to 950 pounds — much too small for the desired show weight.

On the other end of the scale a frame size seven may not fit your desired show weight either, unless you feel a 1,350 or 1,400 pound steer is OK for that show or unless you are going to shrink this steer back to fit a desired market weight. This is a practice I really don't recommend for junior showmen.

If we say a slaughter steer should weigh 1,150 to 1,250, the chart would say we are looking for about a frame size five.

Those with experience can make this estimate by the eyeball method, but using the chart and taking some measurements can be a good educational exercise for 4-H and FFA members. It doesn't hurt to recalibrate the older eyeball occasionally, either.

How old should the steer be? First he has to be old enough to meet the desired weight. This generally means a steer between 15 and 22 months of age at showtime. Animals in the older part of this range are more likely to produce the necessary marbling in the ribeye muscle to meet the choice grade, but those in the lower part of the age range are also capable of grading if fed properly.

Let's say, for example, that the calf is 12 months of age, has a hip height of 49 inches, and weighs 800 pounds six months before the show. If we want him to weigh 1,200 pounds at the show, he has to gain about 400 pounds over those six months.

We can expect the steer to gain about 1.5 pounds per day on a lower energy growing ration of good quality hay and four or five pounds of grain per day. So, if we feed the growing ration for the first 60 days, the steer gains about 90 pounds during this period.

Many feeders would have the steer on a high energy growing ration at least the last 120 days before the show and expect him to gain 2.5 to 3 pounds per day on that ration. This means that the steer would gain about 300 pounds during the 120 day finishing phase.

Most judges are looking for a steer weighing from about 1,100 to 1,300 pounds.

From these figures, you can see I have selected a 12 month old, frame size five steer, and fed him to a perfect 1,190 pound show weight in six months. And I did it without even leaving the office. It gets somewhat more complicated when real animals are involved.

If we wanted to carry this calculation one step further, we might feed this information into a computer and let the computer make the prediction. This would give the same answer, of course; but then we would have it on computer paper.

This judge has his work cut out for him.

7

Feeding Market Steers

There is truth to the old adage, "The eye of the master fattens the cattle." Experience is an important ingredient in the rations of top showmen.

On the other hand, the eye of the judge places the steers in the show ring, and today's judges would encourage the master to go easy on the fat. The 4–H or FFA member should be looking for a practical feeding program that will bring his steer to market weight and show condition. After that, a good steer (and maybe a good fitting job) will have to do the rest.

If you are new to the junior steer feeding game, you'll receive plenty of advice on how to feed your steer; and you may sometimes be surprised by the inconsistencies in recommendations. However, if we can keep an eye on the basics of steer nutrition, this advice doesn't seem so contradictory.

Experienced showmen generally prefer simple, practical rations. If they toss in a cup of "Grandma's Hair Raiser" or a glug of molasses, they do it mostly for fun. When this showman places at or near the top, he knows he had a good steer and a good basic feeding system. While the inexperienced may think it was "Grandma's Hair Raiser" that tipped the scales, we know better, don't we?

Now, I promised to tell you how to feed a steer, didn't I? Let's begin about eight to ten months before the show this steer plans to attend.

First of all, let's get the steer insured. Not that I am nervous about this good feeding advice I'll be giving you; but steers die at the most inopportune times, which is usually when they aren't insured.

Insurance for 4–H and FFA livestock projects may not be available in all areas. Grange Insurance provides this coverage in Washington. In states where this coverage is available, 4–H and FFA members can protect the large investment they have in a steer for only a few dollars in premiums.

After the steers are insured we should plan to feed them at regular times each day, such as 8:00 A.M. and 6:00 P.M. Something near a 12 hour interval is best. Obviously, a feeding schedule like 10:00 A.M. and 4:00 P.M. is going to cause digestion problems. Fresh water and trace mineralized salt should be available at all times.

It is normally advisable to feed two or more steers in the same pen because they eat better this way than when one is fed alone.

Now, let's help this steer set some goals for himself. As discussed in the section on selection, the steer should weigh about 1,100 to 1,300 pounds at show time. (Check locally to see what the grand champion steers have weighed the last few

Weigh the steer as often as possible during the feeding period.

years.) In today's market, we want the animal to grade choice at slaughter.

To determine the length of feeding period and proper show weight for a particular steer, we need to consider the age and frame size of the animal. We can thereby come up with an estimate of the slaughter weight this critter should achieve to have a reasonable amount of fat cover (and marbling within the muscle) to place well in the show and have a good chance to grade choice. This was also discussed under "Selecting Steer Projects."

The major factor in reaching the choice grade is the amount of fat marbling in the ribeye muscle. One of the ways the judge estimates this internal marbling is by looking and feeling for fat covering over the ribs of the animal. There are several factors involved, but most judges estimate the modern steer needs a minimum of 0.35 to 0.40 inches of fat over the upper part of the ribs, before the animal is likely to have the required fat deposition (marbling) in the ribeye to make the choice grade.

There are several ways the feeder can bring the steer to this point. The most common practice is to feed a high fiber, low energy growing ration during the early part of the feeding period up until about four months before the show. At that point the steer is gradually converted to a high energy finishing ration.

Oh, I do hate to tell you that! Some steers may need 140 days on a finishing ration and some need 100 days. Some showmen may feed the same ration for the last six months before the show. We do have to start somewhere, however; and the change from growing ration to finishing ration four months before the show is probably the most common practice for the management systems and project requirements in many areas.

A growing ration includes more hay and less grain than the finishing ration, making this feed higher in fiber and lower in energy. Many feeders also prefer a higher fiber concentrate, such as oats or beet pulp for growing rations. Cost and availability of particular grains should be a prime consideration.

A practical growing ration used by some junior steer-feeders in the Northwest is a grain mix of ⅓ barley, ⅓ wheat, and ⅓ oats, fed at the rate of about four to six pounds per day, along with all of the alfalfa hay the animals will eat. A steer can be expected to gain about 1.5 pounds per day on this ration. Corn could be substituted for the wheat, and other grains could be substituted for barley in areas where this is not a common feed.

A growing ration for steers should contain about 14 percent protein for the total ration, including hay as well as the grain concentrate.

If grass hay or poorer quality alfalfa is fed, a protein concentrate will be needed to provide adequate protein for younger steers. Many junior livestock members may prefer to buy a prepared grain mix, and this may be the most economical alternative in some situations.

Don't guess at the weight of your feeds. Use the scales.

Many junior, steer feeders prefer a good quality grass hay or a grass-alfalfa mixed hay for the finishing period. There may be less chance of bloat or digestive problems with the grass hay in the mix rather than straight alfalfa.

About 120 days before show we'll want this steer on a high energy finishing ration. The steer is converted to this ration gradually, by decreasing the amount of hay in the ration and increasing the grain concentrate portion by ½ pound or less per day until the animal is eating only three to five pounds of hay per day and is receiving all of the concentrate he will clean up.

A finishing ration should contain 12 percent protein. For those who prefer to mix their own grain concentrate rather than buying a bagged mix, the following is a commonly recommended finishing ration:

Wheat	25%
Barley	50%
Oats	10%
Beet Pulp	10%
32% Protein Supplement	4%
Trace Mineralized Salt	1%

Many variations of this ration are possible, depending upon grains available and current prices. Corn and oats will be more available in many areas than wheat, barley, or beet pulp. Check your 4-H books, and talk to experienced feeders to learn about recommended ration alternatives.

If we estimate the steer will gain 1.5 pounds per day on the growing ration and about 2.5 to 3 pounds per day on the finishing ration, we can estimate about what size the steer should be at a particular date before the show. We can also adjust the ration or the two feeding periods to put the steer at the desired weight and finish at show time.

Of course, we have ways of checking up on a steer during the feeding period. It's a good idea to weigh him once a month if possible.

Market Steer Economics

Market steers are the glamour project of junior livestock shows. They're big and look nice when fitted for show, and they produce a big fat check at sale time. We should also remember they produce big fat bills for several months before the show.

The trend at many junior livestock shows has been away from market steer projects and toward smaller animals, such as pigs or lambs. Much of this change is in response to the costs of a steer project and prices received at market livestock sales.

The profitability of a steer varies greatly from one show to another, depending upon sale prices. It's not uncommon, however, to end up with a sale check that just about covers expenses.

A county agent friend once asked, "Who says a kid should make money in 4-H? If the kid joins Boy Scouts, we don't expect him to make a profit, do we?"

The philosophy behind that question says, "Let's keep it in perspective." Because economics is a part of the project, it's nice for the kids to make some money; but profit is not our only objective.

From the junior-sale buyer's viewpoint, he's just looking for a way to support the kids; and he knows about how far his money will go. The buyer looks at the 1,200 pound steer at 20 cents over market price (the buyer's cost if he sends the animal to market) and says, "That's $240 this critter will cost

13

me.'' Then he looks at the 100 pound lamb selling at $2.40 per pound over market and says, ''That's still $240.''

If there's more profit in feeding a lamb or a pig than for raising a steer, many youngsters and their parents will prefer the lower investment and decreased risk of the smaller animals. Facility and equipment expenses are also less for the small animals.

I shouldn't be accused of prejudice. My kids raise lambs, but I believe the steer raiser deserves more profit for his project because of the higher risk and investment involved.

Market steer feeders don't have it so tough everywhere. There are many junior sales where the buyers don't even look at the sale sheet. They just write out a big check and say, ''Give it to that cute, little girl in the cowboy hat.''

Some folks see junior livestock sales as unfair and go to great effort to correct inequities, but I advocate caution when tinkering with sale rules. A lot of good enthusiastic support can be lost when we start ''fixing things that aren't broke.''

Steer shows are fun and educational.

We may find other ways to improve the economics for the steer project member. The kids aren't just in it for the money, but they should know where the money goes and consider how they might get it back.

I remember the day 25 years ago when my vo-ag teacher nearly had a heart attack over prices paid for FFA steers. A couple of our members had gone out and bought four of the fanciest little Hereford steers you ever saw for 40 cents a pound. The current market was about 22 cents a pound. They were hoping for the grand champion but apparently hadn't considered that all four calves couldn't be the champion.

One of the boys said, "I know it's a gamble. My dad says farming is just a big gamble anyway."

The ag teacher took issue. "What do you mean a big gamble? If you want to gamble, you can roll dice, deal the cards, go to Las Vegas! We're not teaching you how to gamble!" He was excited.

After everything calmed down, those two boys raised the steers, placed them near the top of their class at the fair, sold them, and lost about $150 per animal. They argued that they didn't really lose any money because Dad already had the corn.

One boy took his "profit" and bought a motor scooter. The second one may have taken a trip to Las Vegas. I can't remember.

Let's look at some of the options that help keep the steer project affordable. First we should purchase a steer at a reasonable price and insure the animal to eliminate some of the risk.

In some states, Grange Insurance will insure 4-H and FFA projects at a very economical rate. I suspect there are other companies around the country who have similar insurance available.

The steer itself is a major item of expense, unless Grandpa gives you one. In this case the cost is still there, but we just transfer it to Grandpa. Either way, the initial cost or value of the steer is largely dependent upon the current price per pound and the size of the animal when purchased.

Some people like to buy market steers early because the beginning cost is less for a smaller animal. This requires feeding the steer longer, but feed expense is spread over several months. It is sort of like putting money into a savings account to be returned (with interest) when the animal is sold.

Buying the steer early also gives the member a smaller animal to work with and a longer period for taming the critter down. County "steer scrambles" and project selection deadlines also help determine time of selection and price in many areas.

There are some advantages to waiting and purchasing a bigger steer, but the problems of taming and breaking a larger steer in a shorter period of time can be serious if things don't go quite right. I don't know about you, but around my house things hardly ever go quite right.

What is a steer project going to cost? Recognizing the many variables involved, let's say you pay a few cents over market price for the pick of the steer calves on the farm; and let's say that's $.66 per pound at current prices.

So you buy the steer early; and it weighs 650 pounds, about eight or nine months before the show. At $.66 per pound, that's $429 for an initial purchase price.

This critter will eat about a ton of hay and over a ton of grain mixture (concentrate) by show time, at which point we hope he weighs about 1200 pounds. If the hay costs $90 per ton and the concentrate mixture about $140 per ton (these prices will be different when you read this), that's about $230 for feed. Let's toss in the salt and other miscellaneous items for an even $250 in feed costs.

At this point we have about $680 in the steer and may have a few other costs for equipment, vet bills, etc. The steer is likely to have cost over $700 by show time. (I see 4–H steer expense sheets of over $800 at our county fair.)

How much does a steer sell for at your show? I don't know, but I would find out if I were to buy one for a project.

Steers Just Wanna Have Fun

It was such a relief to see the black steer coming down the road dragging a rope halter behind him. I had never been to the farm before and was sure I must be lost, but now I knew I had arrived at the 4-H steer pre-show.

This was spring weigh-day for steer projects, and I could see it would be another one of those days. It was our first spring weigh-day in that county, and we learned a few things.

I jumped out of my car to head off the black steer, and he made a right turn into the pasture with three boys in hot pursuit. The boy who owned the steer got close enough to grab the lead rope just as the animal broke into a run.

The showman dug his heels into the wet pasture grass; and the steer changed leads, but he didn't slow down. The youngster hung on like a bulldog, while the other 4-H'ers ran along behind shouting encouragement (to the boy, I think).

It was quite a show, sort of like a scoop shovel race without the shovel. It had rained all morning; the grass was wet; and the youngster didn't mind being dragged around the pasture like a human anchor. Finally the steer got tired and was subdued.

It was a small show with only eight steers to weigh, and the steer in the pasture was tame compared to the one that got away. I don't mean the one that got loose; I mean got away!

The one that got away was last seen going down the road in a westerly direction and was finally corralled in a neighboring township a couple of months later.

This steer arrived at the fair several months later with two twenty-foot log chains fastened to his halter, and he got loose again. But this time he didn't get away. The kids caught him a half-mile from the grounds when his chains became wrapped around a log.

It would be a shame to have these experiences without learning something. There are several lessons here, I think.

First, it probably wasn't the kids' fault or the steers' fault. We just weren't prepared for handling these animals. In the case of the steer that took a two month vacation, it's hard to break one of these critters after he's been to town and seen the bright lights, so to speak.

A second lesson is to beware the steer with log chains around his neck.

Let's talk about some methods for breaking a steer to lead and keeping him that way. And let's remember that even the apparently gentle animal can cause problems when put in strange surroundings.

Breaking a steer for a junior show begins with selection. A steer with a bad temperament can be a disastrous experience for a junior showman. No matter how good its conformation may be; it isn't good enough.

It's best to begin with a steer of good temperament.

Sometimes a steer doesn't show a temperament problem until you pen him and try to teach him to lead. This is one of the main reasons for beginning to work with a steer early. If the animal can't be handled, it's best to get another one.

In general, the smaller the steer and the more time you have to work with the animal, the easier he is to train. If you happen to get an unruly one, there may be time to replace him with a more suitable steer.

Start training them early and when training the steer to lead, don't put him in a position where he can get away. This means if your county has a weigh-day, the steer needs to be broken to lead before then.

Most showmen recommend haltering and tying the steer as the first step in breaking. Some tie them overnight, some for a few days, and some for only an hour or two in the beginning. The best place for tying is next to a barn or a solid wall where the animal can't get a foot through the fence and injure itself.

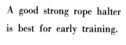

A good strong rope halter is best for early training.

19

Putting the halter on the animal, tying him up, and taking the halter off will help gentle the steer; and having the animal used to the halter is more important than tying for a particular period of time.

A strong rope halter that tightens under the animal's chin when pulled is best for teaching the animal to lead. Leather or nylon halters that don't tighten or exert pressure are no good for either tying or leading a large steer. Of course, your show halter is leather and has a chain to fit under the animal's chin; but you don't want to tie with this or to use it for any of the little rodeos that can develop when breaking the steer.

A strong pen and a strong halter will prevent a lot of serious problems and keep the steer from practicing his escape routine. Some people suggest you begin by leading the steer to water or feed on the theory that the animal wants to go there.

I would suggest the water and feed be in the secure pen, just in case the animal isn't thirsty at the moment.

Many steers have learned to lead after being tied to the back of a tractor or truck and led around in this way. Safety precautions regarding how the animal is tied and where it can put its feet are important, of course.

Some steers learn that an 80-pound 4-H member is fun to butt around; and although they are not mean or wild, they give the kids fits. An equalizer in the form of a short broom stick is generally recommended, and should be applied gently to the bridge of the steer's nose each time he becomes rude.

This should be used carefully, only when the steer butts and in the manner of training while leading, rather than to settle a long-standing grudge. We would like for the steer to associate the broom stick with his bad habits instead of with the kid who feeds him. Also, the bridge of the nose is a tender spot, and we don't want to see the animal injured.

It's sometimes interesting to note that the wildest steers often belong to the older junior showmen. I think this shows that it's not how big you are that's important but how much time you spend training the steer.

Fitting Market Steers For Show

Fitting and showing is one of the most interesting and exciting elements of 4-H and FFA steer projects. Many youngsters would learn far less about their project animal if these activities were not a part of the competition.

While most beginning showmen have older members, leaders, or parents to help and advise on showmanship techniques, there is still plenty of room for panic when one considers the scarcity of printed information on this subject. Many of the accepted practices are unwritten, and those skilled at fitting steers have generally developed their craft through experience, rather than reading about it.

There is hope for the beginner, however. Attendance at a few steer shows, combined with close observation and asking a few questions will generally provide enough information to make the newcomer competitive.

Modern steer fitting methods generally require more clipping than was common in the past, which has a tendency to take much of the fitting job out of the hands of the youngsters. Only the older and most advanced members (or adults) can do what's needed in a professional style clipping job. This may be unavoidable, but we should be conscious of who's doing the work.

This isn't unusual, though; we do the same sort of thing when we clip bellies on lambs or sometimes shear the entire animal. I think we can compromise by letting the kids do all they can and then give them some assistance with things that can be dangerous for a youngster.

If we let the young showmen do as much as possible, they will develop the experience needed to do an expert job in later years. Most judges are aware of the situation and don't expect a professional looking job from a younger showman.

With that said, let's look at a few of the basic recommendations for fitting junior show steers.

Clipping the belly, brisket, head, and at least part of the neck has become almost a requirement in most areas. The modern steer is supposed to look trim, and clipping of these areas is designed to give that effect.

There is no set rule as to where to stop clipping around the steer's head and neck. It is common practice to clip the head to a point several inches behind the ear (some say one clipper width behind the ear) and then blend this down into the lower neck and brisket.

Some clip the outside of the ear, while others suggest no clipping on the ears, or thinning only. The inside of the ear should not be clipped.

The entire brisket and belly is clipped and blended in with the rest of the body. The experts may also clip areas on the shoulder, legs, and topline to give particular effects or blend

Clipping bellies and trimming hooves is much easier with this type of equipment.

You are never too young to learn.

areas together. It's best to have advice from an experienced showman before getting too active with the clippers.

Styles have changed in clipping of the tail and tailhead. We used to clip most of the tail from the twist area up, but it is now common to clip only a small portion of the tail and to leave longer hair around the tailhead. Hair around the tailhead is clipped and combed to help level out the topline and to give the appearance of body length.

Tail switches are commonly formed into a ball by combing, spraying with commercial preparations, and tying up the switch to the shaft of the tail. Many of these things are related to current styles, but some also have a practical value for improving the appearance of the animal. The short, puffy tail jobs help to make a steer look taller; and some say that if you pull the tail switch up too much the steer will look heavy-fronted, which is undesirable.

Some of the best printed matter I've seen on the subject of clipping is an article in the March, 1985, issue of the *Angus Journal*. The article was written by Wendy Gauld of Santa

23

Ynez, California. Copyright laws prohibit copying of such articles, but interested persons may be able to obtain a copy or information from the *Angus Journal* or Wendy Gauld.

The American Hereford Association also has a slide set and script on fitting and showing beef cattle. This is available to 4-H clubs and FFA chapters. Some county extension offices or universities have purchased copies of this slide set.

Steers should have hooves trimmed two to four weeks before show to allow for sore spots to heal, if any are created.

Dads, Moms, and leaders are needed for some jobs.

We still have plenty of washing, brushing, and combing to do for the steer to look his best at show time; and no one will feel comfortable until we spray something on the hair of these steers the day of the show.

Everyone has a favorite product to spray on a steer, but it's important to use something that dries well and is not oily because most judges have an aversion to oily steers. Some people spray the hair with just plain old water and do quite well.

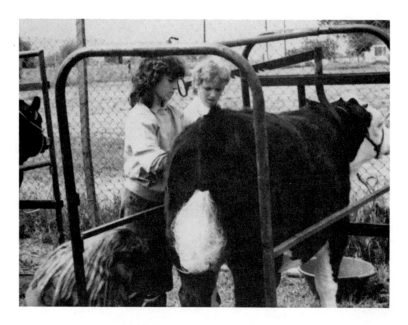

Working together is part of the fun.

Be sure to check the show requirements before using sprays or commercial preparations on steers. Some junior shows have rules forbidding the use of any sprays and waxes.

Use of commercial preparations can certainly be overdone. I have seen instances where the kids used so much sticky goop that the steers collected dust and dirt all the way to the ring.

Then let's make sure the show halter fits the steer and the lead strap comes out of the left side of the halter. The steer is led from the left side. The fitting job won't help much if the animal isn't well trained and under control.

One of the best ways to learn about techniques for fitting and showing cattle is to attend shows and talk to the showmen. Breed magazines also are a good source of information and often contain pictures of well fitted animals.

I think most county agents, 4-H leaders, and Vo-Ag teachers would plead for moderation in fitting junior livestock projects. As techniques for fitting animals become more elaborate, the adults seem to get more involved in the activities. This is not always a good thing.

25

Steer Judges Can't Afford to be Human

Of course they are only human, but I don't think we should let them off with an easy excuse like that. We hire them to judge steers! If they want to be human, let them do it on their own time. Anyone can be human; only a select few can judge steers.

If it weren't for carcass data, steer judges wouldn't have to be human. Years ago we had judges who could line-up a set of steers so quickly they were a marvel to watch. These judges would speak fondly of the smooth and mellow steers at the top of the line and drive triumphantly out the gate in their Lincoln Continentals, smiling to everyone as they passed. The steers went out the back gate in a truck, and everyone was happy as could be.

Now, however, we have carcass data, and many of the old-time judges became human, while others gave up the game to the young and courageous. Nothing is more destructive to a judge's self-confidence than a set of carcass data.

So into this dank and gloomy pit steps the steer judge, who now drives a Datsun and leaves the motor running. The judge endeavors to place the steers in such a way that each animal in the line is slightly better than the one below him, and somewhat inferior to the one above. The judge is able to do this by virtue of his keen eye, superior mental discipline, and undisputed possession of more guts than anyone else on the fairgrounds.

The judge never fails this test. He always gets the steers lined-up, somehow, and picks for the grand champion a clean and lovely beast, considered to be a vision of bovine perfection. Then he jumps in the Datsun and goes for the gate.

Let's be fair and give the judge credit for lining up the steers. He could have just left them milling around the arena, to be rounded-up later by the sale committee. I'm sure many judges have considered this option.

So now the carcass data arrives, and look at what happens. The grand champion grades "good" instead of "choice"; the reserve champion goes "standard," and of 48 steers only 12 grade better than "good." How can we explain this sad state of affairs?

I'm not sure I can explain it, but I'm going to try. Not being a steer judge myself, I'll only attempt to explain what the judge is up against and hope I don't offend anyone in the process.

Let's start with why there are instances when only 25 to 50 percent of the steers in a junior show may grade choice, while we are told that some feedlots expect 80 percent of their cattle to grade choice. Most of the reasons for this are obvious: the show steers are fed differently, hassled the last few weeks before fair, hauled to the fairgrounds and mixed with strange cattle. After all this they are led around the show ring and thoroughly humiliated by some guy with a microphone. Then they are hauled to slaughter with a strange bunch of pampered and pestered pets.

That should explain part of the problem. Many of the things we do in exhibiting and showing cattle cause stresses, which can affect the amount of fat deposition within the meat (marbling) and thus cause the cattle to receive a lower carcass quality grade.

The judge, on the other hand, tries to estimate the fat covering over the back and ribs of a steer, as well as deposits of fat in other parts of the body, as indicators of finish and marbling. These factors are part of what he uses in his prediction of carcass grade.

However, the judge also knows there are other factors affecting marbling, including heredity, age and maturity of the animal, and various aspects of the feeding program. All of this knowledge probably makes the judge nervous, but he tries not to let it show.

These Hereford steers should grade choice. But only a few
people will remember if they did.

We also have asked the judge to place the steers according
to their desirability to the producer and feeder, as well as the
packer. This means he must place cattle that would gain well
and be economical to raise, as well as producing a desirable
carcass. Toss in the fact that a trim "good" grade carcass is
sometimes more valuable to the retailer than an overfat
"choice" carcass, and we can see the judge has been asked to
do more than just assure that the grand champion grades
choice.

How does a judge look at the concern for having a champion that will grade choice? "I look for a steer to have at least
three-tenths of an inch of fat cover over the ribeye to assure
that the steer has a reasonable chance of grading choice,"
states Leath Andrews, Okanogan County Extension Agent
and beef cattle judge. "Beyond this, there is always a certain
amount of risk that the champion will not grade. The type of
feeding program the steer has been on is an important factor
influencing carcass grade."

The author's personal experience with the Klickitat County, Washington, Junior Fair steer show will help explain what
a judge faces. For three years running, we had a contest for

the show audience in which four to six steers were brought into the ring, and the audience estimated the carcass quality grade and yield grade for each animal. Many fairs have done this.

The first year the contest was won by a nice lady, who was quick to admit her previous steer judging was done at the meat counter of the local grocery. Everyone who knew anything about cattle missed them by a mile.

We figured this was a fluke. The second year, the contest was won by a rancher who knew something about cattle. However, it's only fair to say that he also missed them a mile; and everyone else missed them two miles. The third year was the same. No one really could come close to predicting how the steers would grade on the rail.

I believe this contest proved (at least to me) that most of us are much better judges if no one keeps a record of what we said. We also learned that it's very difficult to say whether a steer will grade choice, given only the information available to the judge.

I also recall a district junior livestock show which obtained one of the top steer judges in the country and flew him in from California to judge the show.

He did a terrific job. His placings made sense; he gave good reasons; and best of all, nearly all of his top steers graded choice. Of course, most of the other steers in the show graded choice, also.

Everyone was happy to have found such a good judge, so the show committee invited this man back the following year. Again he did a terrific job. Good placings, good reasons; but many of his top steers didn't grade choice. Some steers graded standard. That was also true of the majority of the steers in the show, by the way.

It was surprising the fellow could lose his judging ability in only one year. He wasn't an old man, either, but apparently his eyes had gone bad; or his mind had deteriorated to the point that he no longer could identify the choice steers.

At least that was the consensus at the next show committee meeting. This judge was never invited back.

Predicting that a junior fair steer will grade choice has always been a risky venture. There are only a few people in the country who can make this prediction in judging with consistency, and most of them have only done it once.

We can always look back and find reasons why some of the steers didn't grade; but when the judge has to predict these things in advance and someone writes down what he said, he is often thankful to see those people only once a year.

Most judges would list the feeding program as the most important factor in determining whether a junior show steer will grade choice. This includes both the length of the feeding period and the quality of the ration consumed.

Beyond the feeding program, most experts list heredity and the physiological age of the animal as the next most important factors. Then we get into some of the finer points, such as how the animal was handled and fed toward the end of the feeding period and how much stress was imposed upon the steer.

There is no evidence that placing a grand champion blanket on the back of a steer will destroy marbling in the ribeye muscle, although one would have to wonder sometimes.

Selecting the Beef Heifer

Selection of breeding stock can be a difficult decision for any livestock producer, but the choice becomes even more critical for 4-H or FFA members who are buying their first breeding animals. An error in selection of market animals corrects itself within a few months, but a mistake in breeding stock selection can be a long-term problem.

This section will present some ideas for selection of beef breeding heifers for a junior livestock member or a small scale producer who isn't already in the beef cattle business. Some suggestions will seem obvious, but others may help prevent serious mistakes.

For example, if you live on a farm or ranch with a good herd of purebred cattle, the question of breed selection may have an obvious answer. Unless you have a pretty good reason to change, you will likely do best with the breed already available at home.

If there isn't a herd of cattle at home, the junior livestock producer has many breeds from which to choose, and much of the choice depends upon personal preference plus what is available in the area. I will only repeat the old statement that the good animals of any breed are better than the poor animals of any other breed.

Importation of many exotic breeds into the U.S. has made us more breed conscious. In some cases, I'm afraid, the emphasis on particular breeds has a tendency to convince folks who are new to the cattle business that some breeds are so superior that even the poor animals are outstanding.

This is not the case of course, and all breeds contain many bulls that should have been steers and numerous cows that would perform better as hamburger.

Once we have decided to buy good animals of whatever breed we choose, we need to decide if we really want purebred (registered) animals or whether grade or crossbred cattle may fit the purpose we have in mind. There are advantages to each; but if the animals are to be shown, we must consider that many shows require breeding animals to be registered.

What should we look for in a breeding heifer? First we would like to buy from a reputable breeder, someone with good cattle, and someone you can trust. Once you have found these people, you have a good source of advice as well as a good source of cattle. You'll find these breeders through advertisements and by talking with other people in the cattle business.

In selecting the heifer, you should ask about production records. You'll want to know as much as you can about the production from the cow and the bull this heifer is out of. The breeder may have weaning and yearling weights of the sire and dam, as well as these weights for other calves from this cow.

He may have a lot more, or he may have less. Either way, you should be very interested in the production within these bloodlines and the number and growing ability of calves produced by the heifer's dam. If the dam stood fourth in line at the Cow Palace, you want to know if she was in the show or just waiting to buy a ticket. If the breeder has good production records, your choice of a heifer becomes much easier.

You are looking for a heifer with good size (and weight) for her age but not one that is overfat. Research has shown that yearling weight is one of the best indicators of lifetime production potential of a heifer. However, research has also shown that heifers that have been fed a high energy ration when young, or those having a tendency to be overfat, will not produce as well over their lifetime as heifers raised on a less fattening plane of nutrition.

We also want good conformation, correctness in the feet and legs, and a heifer showing good femininity. These characteristics are described in livestock judging manuals and

A nice Hereford heifer with good size for her age.

can be learned through practice judging. Heifers with short heads; short, thick necks; and very heavy muscling lack femininity and are not good breeding prospects. It seems that much of what we used to think was desirable in beef cattle has proven to be detrimental to fertility.

The health program of the herd from which you are selecting should be one of your most important considerations. If the heifer is not healthy or transmits a disease to other cattle you may have at home, this could become a most expensive critter.

Recommended vaccination programs will vary with areas of the country and even within the same state. Many states require brucellosis vaccination for breeding heifers, and this is one of the first things to check when buying a heifer.

Regulations concerning brucellosis vaccination have changed frequently in recent years, and I would suggest discussing them with your veterinarian or state department of agriculture. There are also requirements for movement of cattle between states and variations for cattle of various ages.

Other diseases commonly vaccinated against are the blackleg and enterotoxemia complex and respiratory diseases, such as IBR and BVD. In bred heifers, vaccination for leptospirosis and vibriosis is recommended in my area. The proper timing and type of these vaccines is critical, also.

33

Feeding and Management of the Yearling Beef Heifer

The period between one and two years of age is critical for the proper growth and development of the beef heifer. The nutritional needs of this age heifer are not especially high, but either overfeeding or underfeeding can have a detrimental effect on proper growth and lifetime production.

A good feeding program for the yearling heifer should keep her growing well through puberty and breeding age and maintain this growth up through calving. At calving age we want a well grown heifer, which can produce and suckle a healthy calf. We also want the heifer in adequate condition to ovulate and breed back in order to have her second calf one year after the first.

It's difficult to talk about age and weight for breeding heifers because of the variability in breeds, but we have to begin somewhere. Let's say you have a heifer of an English breed, such as Hereford or Angus, and this heifer weighs about 600 pounds at 12 months of age. Larger breeds and many animals of the English breeds may exceed this weight, but we can use this example.

A yearling heifer such as this needs to continue growing at least one pound per day to breeding age of 13 to 15 months. We would prefer she grow more like 1.2 to 1.5 pounds per day. This would put her at 650 to 700 pounds and growing well at breeding age.

Again, larger breeds and larger type heifers will exceed this weight at breeding age. However, while the yearling heifer needs good growth and size, we don't want her fat.

To obtain 1.2 to 1.5 pounds of gain per day the heifer will need good pasture. A small amount of grain (two to four pounds per day) and protein supplement will be needed if the animal is on poorer quality pasture or other forage.

If good quality alfalfa hay is fed, three or four pounds of barley or two to three pounds of corn per day could be added to the ration to achieve the desired gain. If a grass hay or poorer quality alfalfa is fed, a protein supplement will also be needed.

Good quality pasture is an adequate ration for the yearling heifer and for beef cows nursing calves.

If the heifer was of good size at breeding, she should continue growing at a rate of about 0.5 pounds per day during the first six months of pregnancy and should gain about one pound per day during the last three months of pregnancy. During this last trimester of gestation the fetus will comprise about 0.7 pound per day of the weight gained by the heifer.

Thus, our heifer weighing 700 pounds at breeding age of 14 months weighs about 900 to 950 pounds at calving. At first calving the heifer should be about 80 to 85 percent of her expected mature weight.

Those numbers surely are easy to write down and juggle around, but obviously we don't have this animal gaining 1.5 pounds per day the day before breeding, 0.5 pound per day the following day, and one pound per day six months after that. Cattle are not that exacting, but these numbers help explain the general feeding pattern we have in mind.

A practical summary of these scientific estimates would say the heifer should be on a good growing ration from weaning to breeding, but we don't want her fat. She can stand a lower plane of nutrition after breeding but should continue to grow.

The last three or four months of pregnancy, the heifer needs a better feeding program because the fetus is developing rapidly. A good feeding program at this time also assures that the heifer will be in good condition at calving and will be more likely to cycle and breed back for the next production cycle.

There is a long-standing controversy among cattlemen as to whether heifers fed too well during late gestation will give birth to larger calves and, therefore, have more calving problems. Although most experts would agree that heifers can be fed too well during gestation (leading to overfatness and calving problems), research has shown that inadequate nutrition during this period can lead to weaker heifers, which may have more trouble calving and may not cycle for rebreeding as they should. The calves may also be weaker with less chance for survival.

Trying to reduce calf size by providing inadequate nutrition to the heifer is generally a losing proposition.

Bull selection and breeding can be a special problem for junior livestock producers or anyone with a few animals. Artificial insemination can solve some problems for the small producer but requires good management and ability to detect heat as well as good technician service. The advantages of artificial insemination include the quality of bulls available and a reduction in the spread of disease which can result from moving animals from farm to farm.

The other option is taking the heifer to the bull or bringing the bull to the heifer. Either way, it's important to breed the heifer to a bull that will sire calves which are relatively small at birth.

The owner of the bull may have to advise on the question of birth weight. There are bulls in most breeds that can be expected to sire calves that are too large for heifers.

We should also be concerned about the health program of the farm or ranch providing the bull, as well as the vaccination status of the heifer.

Your veterinarian should be consulted about recommended vaccination programs for your area. Many states require brucellosis vaccination for all breeding stock. In my area this vaccination is required for all breeding heifers, but regulations vary by state and are prone to change on short notice. Your vet can advise on current requirements in your state.

Assisting Cows With Calving

Researchers tell us over 70 percent of calf losses are associated with calving difficulty. Although the person with only a few cows may not need to assist with calving often, he will sometimes have to decide if assistance is needed.

Anyone with a large number of cows has had plenty of experience with calving difficulties; but for the small-time operator, deciding if calving assistance is needed is as important as knowing how to help.

My personal experience with assisting cows in calving is limited to an FFA project many years ago. Therefore, I won't pretend to be an expert. However, since the experts are busy helping with calving and few of them have the time to write about this subject, I'll pass on a few suggestions the experts have made in the past.

I do have the distinction of never owning a cow that would calve unassisted. She had two, and the vet pulled them both.

Let's first acknowledge that a major cause of calving difficulty is the bull. The size of calf in relationship to the size of the cow or heifer is the single most important factor in calving problems.

This is greatly influenced by the bull that sired the calf. We can't blame the breed of the bull because calf size from individual bulls varies greatly within breeds.

One knowledgeable cattle breeder points out that blaming the bull for calving problems is an oversimplification of the problem. The breeder must also be concerned about the pelvic size and development of the heifers produced, as well as the birth weights expected from sires.

Another told me studies have shown bulls that tend to sire small calves also sire heifers which grow up to have small pelvic openings. The breeder needs to be sure he isn't solving one problem by creating another one.

Some years ago several universities did a lot of work on the use of pelvic measurements to help predict future calving difficulties in heifers. They found that the pelvic opening is generally related to the size and age of the heifer, that the pelvic opening doesn't grow at a constant rate, and that there is considerable variability in calf birth weights between and within breeds.

While there is no simple answer to this issue, the beginner who has purchased a heifer or a few heifers would do well to consider the calving difficulty that can result from breeding to the biggest bulls without also considering the calf birth weights expected from these bulls.

Cows and heifers still have calving problems despite everyone's best intentions. In order to know when to assist or when to call the veterinarian for professional help, the owner needs to understand what is normal. Some of the best information I've seen on this subject is in the book *Cow-Calf Management Guide — Cattleman's Library* available from cooperative extension in the western states. I would expect similar books are available from universities in other parts of the country, also.

The normal birth process is a longer story than can be dealt with here, but calving is generally described as three stages. These stages are well outlined in *Cattleman's Library*, and I would have a copy if I owned cattle.

The first stage begins with a general uneasiness and signs of slight pain, such as the cow kicking at her belly. Cows on pasture will seek an isolated place.

During the latter part of this time, uterine contractions become stronger; and straining is evident. The first stage is so variable in length I hesitate to say what's normal. *Cattleman's Library* says, "This stage lasts for an average of two to three hours in a cow and four to five hours in a heifer, although it can continue normally for longer periods."

The second stage of calving is characterized by more serious straining, and the animal becomes nearly oblivious to her surroundings. A lot of things happen during this period, including positioning the calf for birth, entrance of the fetus into the dilated birth canal, rupture of the water bag, and expulsion of the calf through the vulva.

My description is somewhat over-simplified, and quite a few things can go wrong during the second stage. Luckily, most of them don't. There are several dangers of assisting too early, including injury to the cow or to the calf. If things are normal, the natural process is best.

The third stage is expulsion of the afterbirth. The afterbirth is normally expelled within a few hours, but may be retained longer. Retention beyond 24 hours is usually considered abnormal and may require treatment.

Some people say there are four stages to calving — the fourth stage being "call the vet!"

When a person should give assistance or call for help naturally varies with the owner's experience and equipment. I asked one of our veterinarians what he recommends.

Our vet suggests the cow should certainly deliver within four hours after the water bag is presented. If not, she needs some help. The water bag will usually be broken before it is observed.

Or if delivery progresses to the point that a part of the calf is showing (foot, nose, whatever) and then there is no progress for an hour or more, it's time to assist. He also suggests that if the cow goes through three or four hours of vigorous labor without results, she needs assistance.

One of the most common mistakes seen by veterinarians is trying to help too soon, before the needed dilation of the cervix and relaxation of the vagina and vulva have occurred. A second mistake is applying traction and trying to pull, even though all parts of the calf aren't properly presented. A third mistake is waiting too long (24 to 48 hours) before deciding the cow won't do it on her own.

When deciding if assistance is required, a good reference, such as the one mentioned earlier, can help the owner determine if labor and delivery are progressing normally.

SHEEP

Why Sheep?

A farm just doesn't seem complete without a few animals. For many small farms a few sheep are the ideal animals to provide just enough noise and manure to make the place interesting.

Young producers enter the sheep business in many ways and for several reasons. Some are interested in profit and want to raise market lambs or breeding stock for sale. Others like sheep for their small size and ease of handling. Many have made rapid increases in their breeding flock simply because the lambs are so cute that the owner can't stand the thought of selling them for slaughter.

The problem with sheep is they multiply so fast that the owner is under constant pressure to decide which ones to keep and which to sell for slaughter. While the cattle owner has two years to make up his mind what to do with a calf, the shepherd will be up to his haunches in woolies if he doesn't think faster than that.

Despite the mental strain, the factors that make sheep and lambs such a popular 4-H and FFA project also make them an important livestock species for small farms. The sheep's small size and low demand for space, feed, and housing fit well into what many junior livestock producers and part-time farmers have available. In fact, many commercial flocks began as 4-H and FFA projects and became a major farm enterprise after the kids left the farm.

Unfortunately, the owner of a very small flock of sheep faces several of the same problems as those who have a few cows or a few sows. Things like how do you isolate newly purchased animals if you only have one pasture, or how do you manage ewe lambs separately from older ewes when you only have one of each?

While these may be extreme situations, similar problems are faced by many. I personally have an eight-year supply of enzootic abortion vaccine with an expiration date of 9/83 and a three year supply of coccidiosis medication left over from the kids' 4-H projects. (If anyone wants to make a trade, let me know what you have.)

Selection of Breeding Ewes

How we go about selecting sheep breeding stock is often dictated by our reasons for raising sheep. Our goals and reasons for raising sheep determine not only the breed we choose but also guide our selection of individual animals for purchase or retention as breeding stock. Whatever the reasons for raising sheep, the 4-H or FFA member or small flock owner needs to set some goals for the breeding program, if he or she is to make much progress.

The first question in selection of breeding stock is whether the goal is to raise purebred stock for sale to others or to produce market lambs and wool for the commercial market or sale at livestock shows. It is possible to do both, of course, but production of quality purebred stock requires a much greater commitment in time and money than does production of market lambs.

I remember a friend and former 4-H leader who used to say, "If we are going to encourage these kids to buy registered breeding stock, we should provide a better means for them to sell the breeding stock they produce." He was suggesting we have some junior sales for purebred breeding stock to give the youngsters a better chance to market their product.

He was right in two ways, I think. One, that junior production sales for breeding stock might be a good idea; and two, that in many cases we should stop talking the kids into buying purebred stock.

While registered rams are the core of good breeding programs and flock improvement, registered ewes are only necessary for production of purebred breeding stock. You'll notice that I use the words "purebred" and "registered" interchangeably. Animals that are said to be purebred but not registered were called "grade" where I grew up, and I would suppose they still are.

Although some youngsters may want to get into the purebred business and "go for the gold," as they say at the Olympics, others would be better off to stick with commercial ewes and buy quality rams. Experience with a commercial flock can help one determine if the purebred business is the desired goal.

Assistance from a knowledgable sheep producer is very helpful when selecting breeding stock.

I recognize that some of the top purebred sheep breeders of today began as 4-H and FFA members. Students of today are the top producers of tomorrow, and we shouldn't discourage them from the purebred business.

Instead, we can encourage those who are willing to make the commitment necessary to produce quality registered animals by helping them attend livestock shows and sales and talk with breeders. Then they should purchase the best quality breeding stock they can afford and be willing to stick with a long-term program of flock improvement.

If the goal is to raise market lambs for show, we want the conformation and muscling of the meat breeds. In most parts of the country Suffolks and Hampshires are the leaders in market lamb shows. However, several other meat breeds are capable of producing quality market lambs and do so in many individual flocks. Excellent market lambs are also produced from crossbred ewes or ewes of the medium wool breeds which have been mated to meat type rams.

There are advantages to crossbreeding for commercial production. Crossbred ewes are more fertile and more prolific, and crossbred lambs produced from these ewes have a higher survival rate than straightbred ewes and lambs.

How do we select individual animals, be they purebred, crossbred, or grade? First, we would like to see some production records. We would like to know the animal's sire, dam, birthdate, weaning weight, type of birth (twin, single, triplet), and the dam's lifetime production records.

Lifetime production records of the dam and any available progeny records for the sire are the best indicators of the performance we can expect from an individual animal. In the case of registered stock, bloodlines, sale records, and show winnings also play a part in selection.

Production records can help tell us if this stylish Columbia ewe has the desired potential.

47

We used to talk about keeping ewe lambs from sets of twins, but the laws of genetics would say the lifetime production of the dam is a more important consideration than whether a particular lamb was born a twin. Therefore, it is recommended that ewe lambs be selected from ewes that have produced well for their age. This means the lamb born a twin or triplet from a ewe lamb or a yearling is more likely to have the genetics for multiple births than a twin born from an older ewe in her peak production years. On the other hand, we can tell more about a ewe with five years of records than about the ewe in her first production season. This must be taken into consideration.

I emphasize multiple births because university research and producer experience tells us this is the single most important factor for profitability in the sheep business. A more meaningful production figure is pounds of lamb produced from the ewe within a period of time (such as 90 days), but this figure won't amount to much if she doesn't have twins or triplets.

Many sheep producers will tell you they hate triplets, and they have good reason. A bunch of bummers can be a problem in some operations. But it's hard to get all twins from ewes without getting some triplets. Many producers are finding ways to make sets of twins from triplets. None have yet figured out how to do this with a single lamb.

Too many lambs per ewe isn't often a problem in small farm flocks. With good feed, some ewes will raise triplets very well by themselves.

Prolificacy and reproduction rate is of utmost importance in sheep production. Selection for high reproductive rate is especially important in farm flocks where good feed conditions make a high level of production possible. Farm flock producers should expect a lambing percentage well over what we would expect for a range flock.

Wool production is also a major factor in some breeds and a minor one in others. Wool production and quality may also be important in selection.

Then there are a host of other factors to be considered in selecting breeding stock, such as freedom from inherited defects, correct conformation, and the general health of the animal. A good reference for beginning, as well as experienced sheep producers, is the *Sheepman's Production Handbook*, published by S.I.D. Inc., 200 Clayton St., Denver, CO 80206. This publication describes the many defects common to sheep and provides helpful information on all aspects of sheep production.

Care and feeding of ewes can be more important than selection for farm flock owners. Proper care of replacement ewe lambs before breeding age is also important.

It is generally recommended that ewe lambs be bred to lamb at one year of age, where feed resources permit adequate growth for ewe lambs to ovulate at seven to eight months. This is not difficult to accomplish in most farm flocks.

When purchasing breeding stock from others, the buyer must remember the health program is of prime importance. All new breeding stock being introduced to the flock should be maintained separately for observation for three to four weeks. This is sometimes difficult for junior livestock producers but can save considerable grief and expense by preventing the introduction of disease to other breeding stock on the farm.

Feeding Ewes

A feeding program for breeding ewes can be relatively simple. The major concern is to allow for changes in nutrient requirements of the ewe according to her age and stage of production.

Let's begin with feeding of ewe lambs with the goal of breeding them to lamb at one year of age. Although there are some situations where feed resources or lambing dates prevent breeding of ewe lambs, most small flocks have adequate feed available to grow lambs to breeding size and condition to lamb their first year. This requires that the ewe lamb have adequate growth to ovulate and breed at seven to eight months.

It is common practice for many purebred breeders or junior livestock members who show breeding stock to not breed ewe lambs their first year, allowing them to attain more growth as yearlings. This is a decision for the breeder, based upon his situation. In the case of 4–H and FFA members we must recognize that we are sacrificing one year's production and also delaying the question of how productive this ewe may be.

Unless the junior showman is on the purebred show circuit and has special need for delaying breeding, he should be breeding ewes to have their first lambs at about one year of age. Some of the more prolific ewes will have twins at one year of age.

We want a well grown lamb at breeding age, but we should not feed the replacement ewe lamb as we would a market lamb. In cattle it has been shown that creep feeding of young heifers reduces their lifetime production as a cow. It seems likely that ewe lambs can also be fed to grow too rapidly, reducing their later production potential. I haven't seen research to confirm this suspicion, but I would expect many experienced sheep producers are cautious about excess feeding of replacement ewes.

What is overfeeding? The best example I can think of would be when we feed ewe lambs for a market livestock show but then decide to take them home for breeding stock instead.

Nutritive requirements of the ewe vary greatly throughout the year. These requirements are determined by stage of gestation, lactation, or maintenance needs.

If we assume a late summer or fall breeding season, we can outline the ewe's feed requirements from before breeding to early lactation.

It is commonly recommended that ewes be put on good quality feed three to four weeks before breeding and that they be gaining weight just before and during the breeding season.

Well grown ewe lambs should generally be bred their first season, and a high percentage will lamb at one year of age.

Studies have shown that increasing the ewe's condition at this time (called flushing) causes an increase in ovulation and results in a larger lamb crop. However, some studies suggest flushing is of no benefit when ewes are bred in October or November, but only for August and September breeding or late seasons, such as January.

Fair quality pasture is adequate for the non-lactating ewe, but her needs increase in late pregnancy and at lambing time.

While it is important for the ewe to be in good condition at breeding, most authorities suggest it doesn't take much increase in energy to achieve this in a farm flock. In many cases the ewes are already too fat. Some information indicates that worming may be enough to achieve a flushing effect in ewes that are already in good condition.

In the corn belt, sheep producers commonly feed 0.25 to 0.5 pound of corn or other grain to ewes, beginning about two weeks before breeding season for flushing. University specialists generally recommend ewes be put on a fresh pasture, such as orchardgrass, prior to breeding and that legume pastures be avoided at this time. Legumes have been found to contain compounds which can alter hormone production and may affect ovulation. Although some of the clovers are the most likely to cause breeding problems, many recommend that alfalfa pasture be avoided at this time, also.

In the West alfalfa is commonly used as a flushing and breeding pasture, partly because that's often the best thing available at that time of year. Researchers say that alfalfa can cause problems as a breeding pasture under a particular set of growing conditions or when other factors may affect growth of alfalfa; but the occurrence of such problems is very infrequent in the West.

Therefore, producers are often willing to take the risk involved in flushing or breeding on alfalfa pastures. Most specialists say clovers are another story, especially Ladino or white clover. On the other hand, I am told producers in areas such as Oregon's Willamette Valley use perennial white clover quite successfully as a flushing pasture.

The possibilities of bloat from legume pasture is an entirely different problem to be considered, also. Your best sources of advice on such matters are specialists and researchers at your state university who are familiar with local conditions.

Many junior livestock members and owners of small flocks face the problem of having the ewes too fat going into the breeding season and may not need any extra feed for flushing. It is important that the ewes not lose weight just prior to or during the breeding season, however. The specialists I talk with emphasize the importance of handling the sheep to determine their body condition.

During the first 3½ months after breeding, ewes need little more than a maintenance ration, as there is very little fetal development during this period. Such a ration can be provided by average quality hay of 12 to 13 percent protein, at the rate of four to six pounds per ewe per day, depending upon the size and condition of the ewe. Ewe lambs and young ewes that are still growing require a higher quality ration than older ewes.

About 70 percent of fetal growth occurs during the last six weeks of gestation, and this period is critical to ewe nutrition. Ewes bearing a single lamb require about 50 percent more energy, and ewes bearing twins need 75 percent more energy than they did during early gestation. This means more and higher energy feeds are required at this time.

A ewe can only hold so much hay, and 0.5 to one pound of grain per head per day is commonly recommended to increase the energy level of the ration during the last four to six weeks before lambing. Again, this is variable with the condition of the ewe and the quality of the forage.

Most small flock owners flush ewes by feeding grain before breeding. The value of this practice depends upon the condition of the ewe and time of breeding.

Insufficient energy during the late stages of pregnancy can lead to pregnancy toxemia or ketosis, which usually results in the death of the ewe. Ewes carrying twins or triplets are more susceptible to ketosis due to their higher energy requirements.

After lambing and during early lactation, the ewe's nutrient requirements are greatly increased. The energy requirements of the milking ewe are two to three times higher than during the ewe's dry period. Like the high producing cow, the ewe needs some grain if she is going to milk to her full potential. The most important element for good growth in young lambs is milk production of the ewe.

A ewe nursing twins or triplets has a higher nutrient requirement than a ewe nursing a single lamb. A common ration for ewes in early lactation is about five pounds of good quality alfalfa hay and one to two pounds of grain, such as barley, corn, or oats, per day.

We should recognize there is a big difference between corn and oats in energy content, and barley is intermediate between the two. Oats is higher in fiber, but lower in energy and may be safer to feed where overfeeding is a concern.

Let's remember all ewes aren't the same size. Some judgment is needed in deciding how much grain to feed. Ewes have to be put on the grain ration gradually. It is generally recommended the ewe receive a smaller amount of grain the first few days after lambing with the ration increased gradually to the desired amount.

The amount of milk produced by the ewe has dropped off considerably by the time the lambs are eight weeks old. Feed requirements for the ewe are also reduced after this point.

This might be a good place to repeat the caution that older references are sometimes out of date on nutrient requirements for certain classes of livestock. As breeding and management practices have changed, we expect much better performance than we used to. When using a reference book or other printed information, I always look at the date to be sure I'm not using out of date recommendations.

Fitting Breeding Sheep

Market lambs are a popular 4–H and FFA project. They are small and, in most cases, relatively easy to handle; and most kids just naturally like to work with the woolly little things. However, good lambs originate from good breeding stock; and the youngsters will learn more about the sheep business if they can show their breeding stock, too.

Several years ago I received a note from a 4–H leader calling my attention to the lack of printed information on fitting and showing breeding sheep. As a result, I picked up some advice from some experienced sheep showmen and have included it here.

We are fortunate that livestock people are generous with information and encouragement. These people are great about sharing their knowledge and ideas.

Let's take a basic approach to fitting breeding sheep, with the recognition that learning is a continuous process. The beginner will learn a lot from his own experience and by watching experienced showmen. I'll try to provide enough basic information in this chapter to get the youngster to the show ring with a sheep looking like its breed and exhibiting all of the confidence and integrity a sheep can muster.

We'll start with Suffolks because that's the most popular breed of sheep in the U.S., as well as the most common at shows. The Suffolk is termed a meat breed; and wool is not a major factor in judging, as it is in the wool breeds. Black fibers are usually considered undesirable, however. Suffolks are shown in much shorter fleece than are the wool breeds, and fitting efforts concentrate on making the animal look tall, long, trim, and muscular, the same as would be desirable in a meat animal.

Many Suffolk showmen shear ewes and rams in the show string about 60 days before a show. This provides a fleece at show time that is easy to work with and gives the animal the desired trim and muscular appearance. The belly and legs are then shorn again a few days to a week before show. The resulting product should look long, tall, trim, and muscular (if you can get them to do all of these things at once).

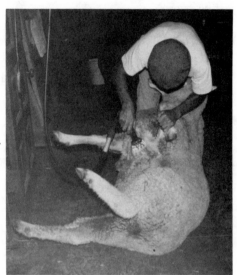

In some areas it is popular to shear the belly of Columbia breeding stock before show, but fitting practices can change. Attending shows and talking to breeders helps one keep current.

The trend in the meat breeds is to shear rather high up the legs as well as shearing the belly and lower part of the brisket. Many Suffolk showmen use small hair clippers on the lower legs and the head to trim up these areas. Any long hair or wool is clipped from the head to give a clean, neat appearance. Hair can be clipped on the back of the ears but should not be clipped inside the ears.

One of the best ways for the kids to determine how they want to fit a ewe or ram is to look at pictures in breed magazines or sheep publications containing sale advertisements and show pictures. Find an animal of your breed that looks the way you would like yours to look, and copy that style.

The late Julie Morris, Centerville, Washington, Suffolk breeder, told me a few years ago that a smooth shearing job about 60 days before show is crucial to a good fitting job. "If the shearing job leaves ridges of wool, the kids will have a very difficult time with final trimming," she said. "It's important to have a clean job of shearing, even if it means going over the sheep a second time to get rid of the ridges." Julie also suggested that the best time for shearing will vary with individual sheep and experience of the showman, but 60 days is a fair average.

Suffolks are washed before show and can be washed right up to show day, as long as they have time to dry and be properly fitted before entering the ring. The wool should always be carded-out well after washing, and should always be wetted for carding. A small spray bottle works fine for wetting the wool. Some folks use sheep dip (disinfectant) in water for spraying the fleece when carding. Others use a water and Borax solution, but some showmen use plain ol' water. I like simple solutions, such as water. (If I don't have to buy it, I'm for it.)

Blanketing after washing and carding is necessary as show day approaches. This helps firm up the fleece, as well as keeping the sheep clean.

What about hooves? They should be trimmed early, about shearing time, and kept trimmed as needed until show day. Hooves should be very clean for the show, but they do not need shoe polish. In fact, shoe polish is not legal in many junior shows. You can see that I have already saved you the expense of sheep dip and shoe polish and have hopes of putting some profit back into the sheep business before I'm through here.

Everything said so far applies to fitting all of the meat breeds, such as Hampshires, Dorsets, Montadales, etc. There may be some minor exceptions, but a look in a breed magazine or conversation with an experienced showman should clear these up.

Now, what do you do to fit the wool breeds? First, we should recognize that all white-faced sheep are not wool breeds. Second, there are three categories of wool breeds: long, medium, and fine. Third, we won't do to a wool breed what we did to those Suffolks. If we do, no one will speak to us.

Attending breed shows or sales is a good way to learn about a particular breed. Gwen Caldwell, Goldendale, Washington, Columbia breeder states, "There are 23 breeds of sheep shown at the Pacific Livestock Exposition in Portland. There's probably nowhere else in the country where you can see so many breeds in one show."

Major breed shows in other parts of the country provide similar opportunities. Sheep magazines and breed associations can also help the beginning showman, if his breed is not plentiful in local shows.

In the medium wool breeds you would like to have 1 ½ to 2 inches of fleece at show time. This means the animals are shorn at least four months before show. You don't want 12 months' fleece on these sheep, partly because it would be very hard to work with, and partly because more than 12 months of fleece is technically not legal in many shows. Therefore, you would normally shear these breeds four to six months before show.

The fleece is trimmed and shaped to give the desired firmness and quality.

Examples of medium wool breeds are Columbia, Corriedale, Panama, and Targhee. Natural colored (black) sheep are generally fitted similar to the medium wools.

In the fine wools and the long wools, you want a very long fleece. I suggest you write the breed association for information on these. Fine wool breeds are Rambouillet and Merino. Long wools are Lincoln, Romney, and Leicester.

For ewe lambs of the medium wool breeds shearing several months before show or an early trimming with hand shears would help get some of the "baby wool" off and make the fleece easier to work with. Columbia breeders have sometimes shorn the sheep's belly, but this can vary according to current customs. The judge may wish to see how far the wool quality carries down on the fleece.

Wool breeds are never completely washed because the crimp and lanolin in the fleece should be maintained. However, all foreign matter, sheep keds, and other impurities should be out of the fleece. The fleece surface should have a clean appearance of uniform color, although it won't be white. When parted, the fleece should be clean and show the desired crimp and quality. Legs and small dirty areas, such as under the legs, can be washed.

When properly carded, the fleece of the dual purpose breeds should be free of vegetation and still show the desired crimp and quality.

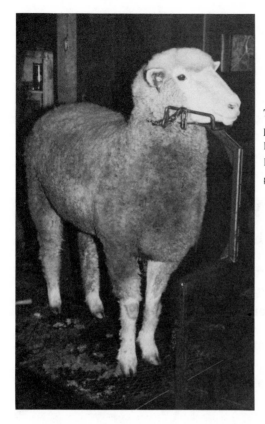

This Columbia ewe is in the process of being fitted. The head has been carded and will be trimmed and blended to give a neat, stylish appearance.

Medium wool breeds are sprayed, carded, trimmed, and blanketed to give the fleece the density, firmness, and quality desired. Trimming is done with hand shears, or electric shears with a blocking comb, following the same principles of conformation as in the meat breeds. The difference is that the quality and amount of wool remaining is an important judging factor.

As mentioned earlier, a very long fleece is desired in the fine and long wool breeds. In the long wool breeds, the fleece is brushed, rather than carded. In the fine wools, a light carding of the tips of the fleece may be possible. I don't plan to go into much detail on these breeds, as I have probably already told you more than I know. Let's write the breed associations; that's what they're for. A list of breed associations and their addresses is contained in the appendix.

Lambing Difficulties

If all went well at breeding, we can look forward to a high percentage of twins (and possibly triplets) at lambing time. How the ewes and lambs are handled at lambing is often the key to success for sheep producers.

Like most other parents of 4-H and FFA members, I am a beginner when it comes to sheep production. I will always be a beginner because I don't ever want to own enough sheep to get much experience.

So I identify with beginners. I know what it's like to pen up a ewe two weeks before she lambs — she sure looked ready to me.

The most difficult lambing problem for beginners is determining what is normal lambing. This decision is further complicated by the difficulty in finding normal sheep. However, there are guidelines that can help greatly in understanding what is going on.

Most land grant universities have printed information on the care of sheep before lambing, at lambing, and after lambing. The University of Idaho has a very good series of leaflets called *Current Information Series No.* 619; they describe management practices at lambing time as well as during other times of the year. I would expect similar pamphlets are distributed through county agents in many other states, also.

If you can't obtain these through your county agent or university contacts, drop me a line. For the cost of the bulletins and postage , I'll send copies to you.

The Idaho leaflet says, "The total time span for normal delivery is about five hours, including four hours for dilation of the cervix and one hour for delivery of the lamb. After a ewe is in hard labor for five to fifteen minutes, the front feet and nose of the lamb should appear."

Let's further consider for us beginners that these times will vary with individual animals and may be longer in ewes lambing for the first time. The pamphlet says normal birth should occur within one-half hour after hard labor starts and the water bag is ruptured.

Signs of lambing difficulty include

    A.Continued straining without appearance of a water bag.

    B.Continued straining for an hour after rupture of the first water bag but without the appearance of the lamb at the vulva.

    C.Partial expulsion of the lamb with the ewe unable to complete delivery.

Once it is decided things aren't progressing normally, you may need to make an internal examination of the ewe. Whether you do this yourself or call the veterinarian for advice or assistance, depends on your level of experience.

Trying to help too soon is a mistake because of the time needed for dilation of the cervix, as mentioned earlier. Unassisted delivery is always best unless a problem arises.

On the other hand, small flock owners shouldn't be surprised if they have to examine some ewes. There will be lots of times when the beginner isn't sure what to do, and sometimes an examination is the only way to find out what is going on.

Cleanliness is of major importance when making internal examinations. The person making the exam should scrub his hands and arms and lubricate with soap or an obstetrical cream. Soap and lubricant are also important even if throwaway plastic gloves are used. The area around the ewe's vulva should also be washed with soap.

Cleanliness, good lubrication, and gentleness are of prime importance when assisting with lambing. As someone said, "Those lambs have been in there five months already; there's no hurry in getting them out now."

Anytime the hand (with or without plastic gloves) is inserted into the vagina, the danger of uterine infection is greatly increased. Most veterinarians recommend insertion of uterine boluses and injections of antibiotics anytime a ewe is examined internally. Specific recommendations and dosages should be obtained from your vet.

I won't attempt to give advice on correcting abnormal presentations. I suggest producers obtain one of the pieces of printed material available from extension offices or universities.

Sheep producers in Washington and Oregon have had the opportunity to get some lambing experience in a hurry by attending one of the lambing schools sponsored the last few years by the Washington Wool Growers and Washington State University. The University of California and possibly several other universities have sponsored similar schools for sheep producers.

Raising Orphan Lambs

Nearly everyone who owns sheep has some experience raising bummer (orphan) lambs; however, most of us are still looking for a better way.

The bummer lamb got his name for his habit of taking a meal wherever he could get it. The bummer is an orphan because his momma died, or she won't claim him, or someone has decided she has more than she can do when it comes to feeding lambs.

There was a time when bummer lambs were considered worthless. For many small sheep producers that time has traditionally been about 3:00 A.M.

The advent of milk replacers and cold-milk feeding systems has made raising bummers more feasible, however; and many shepherds have nearly forgotten the hatred they once felt for those helpless, little creatures. It's still not a lot of fun, but feeding orphan lambs has become more manageable.

As producers strive for higher lambing percentages, the number of lambs that must be raised artificially increases. A good artificial milk feeding system can save many lambs that would otherwise die.

While the economics of feeding bummers on milk replacers is always a subject of debate, I think nearly everyone would agree that bummer lambs can be raised economically at today's prices; but there's not a lot of room for mistakes.

Some sheep operations sell bummers to 4-H or FFA members (or others who want to raise them) at about $10 to $15 each. These are reasonable prices when the cost of milk replacer and other costs for raising orphan lambs are considered.

Research at the U.S. Sheep Experiment Station in 1978, estimates costs of over $15 per head for raising orphans to a weight of 45 pounds using a cold milk feeding system followed by a 39 day post-weaning period on a high quality creep ration. These figures would indicate a bummer purchased for $15 would have a total cost (excluding labor) of about $30 by the time it weighs 45 pounds.

Because these are 1978 figures and are based on high volume commercial production, we can be sure the small producer with a few bummers is going to have more expense. There is also a lot of labor involved.

Whether it's considered possible to raise orphan lambs economically and to produce good show lambs from them depends upon who you talk to. It is definitely less likely that a bummer will develop into a good looking lamb compared to a lamb raised naturally on the ewe. I am told it can be done, however.

In addition to the health problems sometimes encountered with orphans, these lambs also have a tendency to develop more belly then we would like in a show lamb. Cold-milk feeding systems combined with early weaning should help reduce this problem.

There are many ways to raise orphan lambs, and those with experience will have their own ideas and methods. Many prefer to stick with the bottle and warm milk method for small numbers of lambs while others prefer self-feeders with cold milk.

I would suggest anyone raising orphan lambs obtain a copy of U.S.D.A. Bulletin 2207, *Rearing Lambs on Milk Replacer Diets*. This booklet contains information of benefit for whatever system of feeding a person might choose and should be available from local county extension offices.

It is important to give the lamb a good start by making sure it has received some colostrum two or three times during the first 12 hours after birth. Colostrum contains antibodies to fight infections, but the lamb begins losing its ability to use these needed antibodies soon after birth and has lost most of this ability within 12 hours after birth.

Many shepherds keep a supply of frozen colostrum milk on hand for instances when colostrum milk is not available from the ewe. Cow colostrum has also been proven of value for lambs. Colostrum should be thawed at room temperature because antibodies can be destroyed by heating.

According to a news release from the University of Idaho a microwave is OK for thawing colostrum. In the release David Olson, a research veterinarian at the University of Idaho, states that he has found colostrum can be safely thawed in a microwave oven set at 60 percent power.

He says microwave thawing results in quality concentrations three times greater than thawing colostrum by warm water methods. He stirs the mixture as it thaws and cautions producers not to exceed the 60 percent power range.

(Those who don't have a microwave may wish to file this under "Reasons we need one.")

U.S.D.A. Bulletin 2207 contains good information on how much milk to feed lambs of various weights as well as systems for self-feeding cold milk. Even the person with one or two orphan lambs can rig up a very simple cold-milk feeding system if this is preferred.

Owners of small flocks often face the situation when a ewe has triplets or even quadruplets and is content to raise them herself. Many experts will recommend that the ewe be permitted to raise two lambs and that the extras be put on artificial milk.

I think we should recognize that some good milking ewes can raise triplets on their own. With a good feeding program and creep feeding of lambs, I have seen ewes raise a good set of triplets. This may also be possible with quadruplets in isolated instances, but that's asking quite a lot.

On the other hand, if the ewe starts out raising triplets and doesn't produce enough milk, the lambs are in trouble. The shortage of milk generally shows up in poor condition of the lambs at about three weeks of age.

An alternative for the owner with a small number of ewes is to leave all three (or four) lambs with the ewe, if she will claim them, and to supplement some of the lambs on a bottle. This can be done by training one or two lambs to drink from a bottle within a few days of birth and then continuing to feed them some milk replacer to supplement what they receive from the ewe. I am told it's difficult to train older lambs to drink from a bottle, so this training should be done within a few days of birth if possible.

My son helped one ewe raise quadruplets this way and another to raise triplets. The two quadruplets fed supplementally on a bottle consumed 25 pounds of milk replacer and were nearly the same size at weaning as the other two quads which got all of their milk from the ewe. The four quadruplets had a combined weight of 240 pounds at 60 days of age.

Two lambs from a set of quads appreciating a little extra milk from the bottle

The 25 pounds of milk replacer fed these lambs was well worth the expense. The same ewe raised a good set of triplets on her own as a two year old.

Getting lambs on dry feed as soon as possible should be a prime goal in artificial rearing. This reduces the cost of expensive milk replacers and results in a healthier lamb. Lambs can be weaned as early as four or five weeks if they are eating adequate amounts of dry feed at that time.

Lambs should have fresh water, high quality alfalfa hay, and a palatable creep feed at a few days of age. When very young lambs are weaned from the milk replacer diet, a creep ration of 20 to 24 percent protein should be available for the first two weeks. Two weeks after weaning the protein percentage of the creep ration can be dropped to 18 percent.

Feeding lambs beyond this period is similar to feeding naturally reared lambs of the same weight. Some research has shown artificially reared lambs to be more susceptible to internal parasites than lambs raised on ewes. It is recommended these lambs be kept separate from naturally raised lambs.

Selecting Market Lambs

When buying market lambs from another breeder, the junior showman's selection process begins with the question of lamb size and age. The breeder can often be of help in estimating what size lamb you should begin feeding for a particular show date, but it is important to remember that your management and feeding program may be different than the breeder's. Talk these things over before making the decision.

A good market lamb can be expected to gain from 0.5 to 1 pound per day for a 60 to 90 day period after weaning, with good management and a high energy ration. On a good quality pelleted ration, 0.6 to 0.8 pound per day gain is likely. On average pasture or a low energy ration, well under 0.5 pound per day is much more likely. (I'm speaking of weaned lambs here.)

Because feeding and management will be discussed later, I'll assume for our purposes that we will feed a high energy ration and will plan for gains of 0.6 to 0.8 pound per day. From this base, we can begin to adjust for the size lamb you have to work with, the final weight desired for your show, and other bits of feeding advice from guys like Uncle Ralph or me.

Let's then look at a desirable size for a show lamb and demands of the current lamb market. I think it's safe to say, what you hear about demands of the market and what you hear about livestock shows won't always match up exactly. This is partly because what we hear depends upon whom we talk to; and partly because a livestock buyer and a livestock judge have to look at things differently.

For instance, a buyer deals in large numbers and has to buy on averages to a certain extent, whereas a show judge is looking for an animal approaching an ideal. If the judge selects a 130 pound lamb for grand champion, he isn't saying he would like to see all lambs in the show weighing 130 pounds. He's saying, for this particular lamb, 130 pounds is a good weight; and the market would be happy with a bunch of lambs like this one. The market would not be happy with the average lamb fed to weigh 130 pounds. The average may not have the frame to finish properly at that weight.

Most judges are looking for a properly finished lamb weighing about 110 to 130 pounds.

I think, given a good selection, most show judges today would like a lamb that finishes from 110 to 130 pounds. In some areas of the country we might narrow that to 115 to 125 pounds. If we accept the ranges given and you want a lamb that will show at 120 to 125 pounds, you can begin to figure what size he should be 90 days before the show. For example, if the lamb gains 0.6 pound per day, he could weigh 65 to 70 pounds, three months before show. A lamb averaging 0.8 pound per day gain (which is good for a 90 day period) could weigh 50 to 55 pounds three months before show and still come in at 120 to 125 pounds.

Weigh the lamb as often as possible to keep track of its progress. Often lambs won't gain much the last week or two before show and may even lose a few pounds.

I think those are fair numbers, but let's be cautious with the kids who don't have experience feeding lambs, have baseball practice four nights a week, and can seldom understand how the water bucket got dry. It is always easier to hold a lamb back a little than to push one that isn't up to size. The junior livestock member has a tough assignment in feeding projects to a certain size by a certain date. When in doubt, you are better off to start with an animal a few pounds heavier than what you might need under ideal conditions.

I should pass on a comment from a friend and expert in the sheep business. She told me, "Don't expect much gain from lambs that last week or two before Fair, when the kids are busy washing, carding, and working them. If a lamb weighs 115 a week or 10 days before Fair, chances are he'll weigh about the same at show time. The excitement and stress of getting lambs ready for show often seems to prevent additional gain.''

Lambs in the 60 to 70 pound range are usually two to three months old. These lambs will be five to six months of age at show time. Again, there are variables; but these ages provide a starting point.

But guess what! For a spring show, you may have trouble finding lambs to fit these age and weight figures, which means you'll need good lambs (many will beat the numbers I've given); and you may have less room for error in management and feeding programs.

When you buy lambs from others, the best time for weaning is a common concern. Sellers of very young lambs may be concerned about damage to the ewe's udder from early weaning.

A few years ago, Grant County, Washington, extension agent, Ladd Mitchell, had this to say on the subject: "Risk of damage to the ewe's udder should be very slight when weaning at 60 days of age, if proper precautions are taken. Ewes should be placed on the equivalent of a dry ewe ration, two to three days before weaning, to help slow down milk production. This would be about three pounds of hay or low quality roughage per ewe, per day, and no grain.''

Ewes should be taken from the lambs and kept off feed and water for 24 hours after weaning. Of course, lambs must be eating creep feed well at this time, and must be kept on the same creep ration for a period after weaning, to avoid any setback for the lambs.''

This should provide a basis for selection of a 4–H or FFA market lamb. You are also looking for good conformation and a lamb with potential for the length and scale needed in a good show lamb. Muscling in the leg and the loin area is also important. Booklets on livestock judging will help describe what to look for in a market lamb project animal.

Raising Market Lambs for Projects

Most junior fair show lambs are purchased at weaning time or later, but a good percentage of 4-H and FFA members raise show lambs from their own ewes. While those who buy lambs after weaning are most concerned about selecting for proper size and conformation, youngsters who raise their lambs from birth have a few more details to consider.

The first and most important consideration is when to breed the ewes to get lambs of the proper size for a particular show. Most lamb showmen have traveled through this wringer a few times and have determined what works for their sheep and their management system. Newcomers, on the other hand, may be justifiably amazed to learn that show lambs of the same weight may vary in age by three months or more.

Market lambs coming to our local fair in September have been born anytime from January to May, giving ages of four months to seven months or more. Breeding and management systems vary, causing some lambs to gain faster than others. With good milking ewes and good feed, it's not uncommon to find that the baby lamb of three months ago now weighs 100 pounds or more.

Unfortunately, the baby lamb of three months ago may now weigh 60 pounds if things haven't gone so well for him. That's what makes it interesting and is one reason why it's difficult to say when lambs should be born. There are a number of variables to consider.

Let's look at some of the variables. A well-bred market lamb, raised as a twin, creep fed, given a good health program and a good finishing ration, could weigh 120 pounds at four months of age. Some breeders achieve this every day.

On the other hand, if we want to hold the lambs back a little near the end of the feeding period or we aren't too sure about the other variables, it's safer to have them about five months old at show time. What time of year the lambs are born could be a factor here, too, especially if pasture for the ewes is a major part of the feeding program.

We could recommend lambs be a little older for members without previous experience; but when we get to lambs of six months or older, a growthy lamb will need a lot of dieting to keep him within proper show weight. Experienced breeders will see I'm trying to cover my backside here. It is easier to hold a bigger lamb back than to push one that just doesn't seem to care about livestock shows.

All of the management factors should be considered. If we expect ewes to lamb five months before the show, we must consider that some may not be bred the first week they are put with the ram.

It's always best to feed at least two lambs together. They like company.

Any youngster feeding a project lamb should have two. Even if only one lamb is to be shown, a second one should be fed, also. They don't do well alone. My kids had four in a pen a few years ago and decided to split them to better control feed consumption. All four lambs ate poorly until they finally put them back together again.

Owners of a few ewes can do a number of things to help get lambs off to a good start. Ewes should be given a booster shot before lambing to prevent enterotoxemia. This will help protect the lambs from enterotoxemia until three to four weeks of age. I suggest beginners talk to their veterinarian for recommendations on timing of these vaccinations. Many growers also vaccinate the lambs at birth, as well as at later ages.

The most important thing for good growth in young lambs is milk production of the ewe. The energy requirements of the milking ewe are two to three times higher than during the ewe's dry period. This means that like the high-producing cow, the ewe needs some grain if she is going to milk to her full potential.

The amount of milk produced by the ewe has dropped off considerably by the time the lambs are eight weeks old. Feed requirements for the ewe are also reduced after this point.

Creep feeding of lambs helps to supplement the ewe's milk and prepares the lambs for weaning with a minimum setback in growth. A creep is simply a place for the lambs to eat without interference from the ewes. This is usually built from panels and is constructed to let the lambs enter, but exclude the ewes. Lambs will begin eating solid feed from a creep at seven to ten days of age if the creep is placed properly.

The creep should be placed near where the ewes are fed or watered or where they gather during the day. Small lambs won't leave the ewes to go to a creep if it is any distance away. The creep must also be open enough so that the lambs can get in and out easily and can see the ewes from inside.

Whole grains or rolled grains are OK for a creep ration for small lambs. Leafy alfalfa hay should also be provided in the creep. The small lamb is receiving a high protein diet from the ewe's milk, and the creep feed makes up a very small part of his ration.

As the lambs begin to consume larger amounts of creep feed and are closer to weaning, the protein content of the creep ration should be increased to 16 to 18 percent. If a prepared ration is purchased, it is important to get one with natural protein rather than a portion of the protein from a urea source. The rumen in young lambs isn't well enough developed to utilize urea protein sources.

Lambs that are eating creep feed well are commonly weaned at 60 to 90 days of age in farm flocks. They can be weaned younger than 60 days under some conditions, but other management factors have to be considered.

Feeding Market Lambs

Life would be much simpler if there was only one way to do things, but it would be so much less interesting. Each time I write something about feeding livestock I worry that there are some things I don't know about the subject, and even worse, some of what I think I know is probably not true.

We can all point to previously held notions about livestock feeding which have been proven false by research. I would expect that a few of our current ideas will also be proven wrong at some time in the future. We are probably all guilty of giving advice based upon what we were told by someone else. A friend told me recently, "It is surprising, the people who were asking me questions last year are giving me advice this year."

This is especially true when talking about feeding livestock for junior shows. There are many good feeding programs and more than one way to get the job done.

Feeding and animal health programs are a combined package with all types of livestock. These can never really be treated as separate subjects. This becomes especially evident when feeding market lambs.

Attention to all aspects of management and animal health seems even more critical for lambs than for other 4–H or FFA livestock projects. Lambs don't die often, but for most of them once is enough. If you only have two and one dies, that's a big loss.

Before we start feeding market lambs, they will need some shots. Most experts recommend that lambs be vaccinated for enterotoxemia C and D (over-eating disease) when they are

78

put on feed. Many lambs received this vaccination at birth or within the first few weeks of age, but they should have another enterotoxemia vaccination when they go on feed. Your veterinarian or county extension agent can help you determine the number and timing for these vaccinations.

Lambs should be vaccinated for enterotoxemia (overeating disease) before being put on feed.

Lambs should also be wormed at weaning and possibly again at the time we begin feeding them for show, if there is much of an interval between these events. In many areas a Bo–Se injection, to protect against selenium deficiency, is recommended at weaning time, as well as at birth.

Coccidiosis has become a serious problem for lambs in some parts of the country. Your local veterinarian, county extension agent, or livestock specialist can advise on prevention or treatment for this disease, also.

Nearly all junior show lambs must be confined for some period before show as a practical matter. If members are to work with lambs as expected, the animals must be accessible, which usually means confined. It is also easier to assure proper finish on a show lamb if you can control the feed intake.

A common program for show lambs is to feed a dry ration for about 60 days before show, with some variance depending upon the size and condition of the lamb at the beginning of the feeding period. It's difficult to predict how fast a lamb will gain because of the many variables in the care and feeding program and between individual animals.

However, gains often average about ¼ pound per day for weaned lambs on pasture alone, and up to one pound per day with good growthy lambs on a high quality pelleted ration. With a high quality ration and good management, many junior show lambs will average about 0.7 to 0.8 pound per day for a 60 day feeding period.

Self-feeding a complete pelleted ration designed for lambs works well for many showmen.

A lamb will eat about three to four percent of its body weight per day. An 80 pound lamb will eat about three pounds per day; a 100 pounder will eat about four pounds per day.

The most popular lamb ration in many areas is a complete pellet containing 12 to 14 percent crude protein. This ration can be self-fed, meaning feed can be kept before lambs at all times. However, many showmen and 4–H leaders or FFA advisors prefer the ration be hand-fed twice a day. Hand feeding twice a day helps assure that the feed is fresh and gives more opportunity to observe the lambs, in case one is not eating properly. A ration with 12 or 13 percent crude protein is often suggested for larger lambs, but young lambs may need 14 percent crude protein or higher when just coming off a creep ration.

Pelleted rations can be fed as the only ration or with a small quantity of long hay. Some feeders prefer a small amount of long hay (one pound per day or less) be fed with the pelleted ration to aid with rumen function. Long hay is simply hay from a bale or coarsely chopped, as opposed to hay that has been ground or pelleted. I tell my kids to give them a little hay, just to make the lambs and me happy.

Grain and hay rations are also practical for market lambs and may be most economical in situations where pelleted rations are unavailable or much more expensive than available grains. Lambs digest whole grains very well, making processing unnecessary for grains to be hand fed. I like the pelleted rations. But if hay and grain rations are much cheaper than pellets, I start looking for hay and grain. There has to be some economics in this project, too.

Corn, barley, wheat, and oats are all commonly used in lamb rations. Corn and wheat furnish the most energy per pound; oats furnishes the lowest energy but the highest protein of these four grains. Barley is intermediate in energy per

Fresh water and shade are very important, especially during hot weather.

Be sure to weigh grains carefully. Rations must be formulated by weight rather than volume measure.

pound. Corn and wheat can be expected to produce the fastest gains but also require more care in feeding.

Remember to weigh all feeds. A coffee can of grain weighs about as much as a rock — quite variable. If we switch from oats to corn, the amount of feed in a one-can measure is nearly doubled. A change from rolled barley to whole barley increases the amount of feed in a coffee can by about 40 percent.

High grain rations are normally fed with good quality alfalfa hay, but an additional protein supplement may also be needed when a low protein grain, such as corn, exceeds 50 percent of the ration. When optimum gains are desired, the grain portion of the ration will exceed 50 percent; but some research suggests the grain portion of the ration should not exceed 75 percent. An example of such a ration would be three pounds of barley and one pound of high quality alfalfa hay per day. (This ration may need additional protein.)

When grass hay or poor quality alfalfa is fed, a protein supplement, such as soybean oilmeal or a commercially prepared supplement, should be added to the ration. For example, if the grain and hay portion of the ration provide a diet containing 10 percent crude protein and are fed at the rate of four pounds per day, the addition of ¼ pound of 44 percent soybean oilmeal per day would bring the ration up to 12 percent.

Many sheep producers are cautious about suggesting high grain rations for 4–H and FFA lambs. When we get up around 75 percent grain in the ration the danger of digestive problems increases. There's something about only one or two lambs in the pen and a youngster on the feed scoop that makes everyone nervous. Commercial feeders commonly feed much "hotter" rations (more grain), but they know what they are doing. One advantage of the pelleted ration may be a greater margin for error and therefore fewer chances for feeding mistakes.

Sometimes lambs that are too big early in the feeding period need to be held to a low rate of gain for a period of time to prevent them from exceeding the proper show weight. This can be done by providing a lower energy ration with most or all of the ration coming from hay or pasture, but this leads to the large-belly problem.

Some showmen report good results from limit feeding the higher energy diet and holding the lambs' weight through exercise. I have also seen one instance where lambs were kept on a high roughage diet early in the feeding period and fed pellets for the last three weeks before the show. This took the excess bellies off these particular lambs.

We should caution 4–H and FFA members to put lambs on feed gradually and not switch feeds suddenly during the feeding period. If we go to the mill and buy a different feed because they're out of our usual, we can get into serious problems.

Don't forget the variables. If the member or parents don't have lamb feeding experience, it's quite possible for lambs to gain nothing at all for long periods of time. Use the scales often.

This lamb is showing severe bloat symptoms from eating more than his share of the grain. This is a nice time to get acquainted with your veterinarian.

In summary:

1. Pay careful attention to the health program.
2. Put lambs on feed gradually. If lambs have been on creep, any changes in kind or amount of feed also should be made gradually.
3. If hand feeding, feed at regular hours each day.
4. If a complete pellet is purchased, get one that is designed for lambs.
5. The lamb ration should contain 12-14 percent crude protein. Lambs under three months of age should receive a 14 percent protein ration or higher in some cases.
6. Know what your feed costs. Someone is paying for it.
7. Don't switch feeds abruptly.
8. Weigh all feeds.
9. Watch for lambs going off feed, indicated by scours or loss of appetite.
10. Fresh water and trace mineralized salt should be available at all times.

Fitting Market Lambs

Fitting techniques for market lambs have undergone considerable change in the past fifteen years. Each of these changes has caused adjustments in our way of thinking and has brought revisions of printed information. In most cases, revisions of the printed information are a long way behind the thinking.

In many areas we now have a body of printed knowledge which combines the best of the old with the best of the new and which, consequently, contains enough conflicting advice to bring tears to the eyes of the most hardened wool puller. It's difficult to keep literature up-to-date with everything that is happening in the show ring and with some of the fads that are here today and gone tomorrow.

On the other hand, beginners have a right to their few moments of panic when they do as the book says and later learn no one else does it that way.

With that said, I'm going to climb out on a fairly short limb and attempt to describe how junior showmen are fitting market lambs in most areas. Remember, if it doesn't fit, don't wear it. Your show may have different requirements or traditions. Along with all this information I think we need to be flexible in what is expected in fitting of lambs for junior shows. Too much emphasis on the fitting job for younger showmen can result in more adult involvement than we would like.

In some states show lambs are shorn very close to show date (less than 10 days). This eliminates much of the trimming and the issue of how long the fleece should be. Find out your show requirements before doing anything.

Many shows in the Northwest require or suggest shearing market lambs about 60 days before show. This is a good practice even if it's not required. A lamb with 60 days of fleece has enough wool to permit considerable carding and trimming but is much more manageable than one in full fleece. A young lamb that hasn't been shorn is very difficult to do a good trimming job on. Those with experience advise that a good, smooth shearing job will save a lot of extra, difficult work in the final trimming job. (Long wool and medium wool breeding sheep are another subject.)

Shearing lambs about 60 days before the show is a requirement in many shows and makes fitting the lamb much more manageable.

The lamb's feet should be trimmed during the feeding period if needed. The feet may need to be trimmed twice during the month before the show. Ideally, the last trimming should be a week before the show to permit time for healing of any mistakes or sore spots.

Trimming feet before the show is necessary for both market lambs and breeding sheep.

Most showmen shear lambs' bellies within 10 days of the show. This varies with experience of the shearer and the possible need for healing of nicks or cuts. Many also shear the legs of market lambs to a point above the hock and then use the hand shears to blend this in with the body.

Advice of experienced lamb fitters is helpful here. Pictures in breed magazines or other publications will also give an idea of the effect we are trying to create.

If I were a beginner and didn't have electric shears or experienced help, I would clip the belly as close as I could with hand shears and cut the hair on the legs and head with a pair of scissors. I can hear the sheep showmen groan. Hand shears can't do bellies like electric shears can, but we sometimes have to make do.

Nearly everyone washes market lambs as often and as close to the show as needed. This should begin about two to three weeks before the show. Most lambs will need two or three washings before they are clean. Special attention should be given to areas under the legs where dirt collects.

Some experienced showmen prefer to dry the lamb with towels (or whatever) after each washing and then card and trim the wool while the lamb is still damp. Carding while the fleece is damp helps to straighten wool fibers and results in a firmer, smoother trimming job.

When buying your first wool cards be sure to get the ones with curved teeth for carding lambs. Cards with straight teeth are for combing and straightening long wool.

Kids should allow plenty of time for washing, carding, and trimming. The smaller ones especially will tire quickly when they have to card and work the hand shears for a few hours.

It's best to begin three weeks or so before the fair to allow enough time for a good job. I asked some experienced sheep raisers how long it might take a beginning member to get the job done. It seems twenty or more hours of washing, carding and trimming time is not out of reason.

The lamb should be kept in a clean, well-bedded pen after washing. Blanketing the last week or two before the show helps keep lambs clean and firms up the fleece.

This lamb has it all. A good blanket complete with hood.

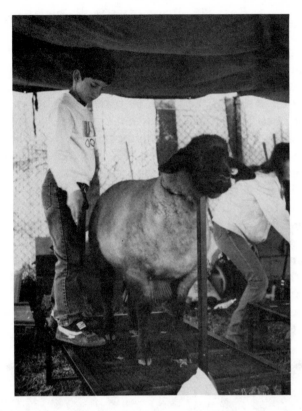

A good trimming stand puts the lamb where you can work on it and keeps it under control.

Some will make a final washing the day before the show. Many shows would prefer lambs be washed a few days before show and spot washed at the show as needed. The animal must be dry by show time.

There often isn't space or facilities for washing all of the animals at a livestock show. Some have rules against any washing at the show.

Fitting stands are very helpful for all types of washing and fitting jobs. Lambs should not be left unattended on the stand, however. Some will jump off the stand and break their neck.

When carding and trimming a dry fleece, it helps to wet the fleece by using a spray bottle containing water or a rag dipped in water. Some people use a small amount of borax in water or just water. Others prefer to add ¼ teaspoon of sheep dip (sold as a type of disinfectant now) to the spray mixture.

Few people use sheep dip for anything else these days, and I wouldn't buy a can for this purpose alone. At ¼ teaspoon per gallon of water a can of dip is an 80 year supply.

How close should the fleece be trimmed? There probably isn't one best length, and I'm afraid a ruler would make liars out of many of us. The most important things are that the fleece is firm and that the lamb feels firm when handled. If the lamb has been shorn 60 days before the show, a common fleece length after trimming in our local shows would be about ¼ to ½ inch. Some shows may recommend the lambs have enough wool to make a number one pelt after trimming. This would require ⅝ to 1 inch of wool. (Talk to showmen in your area.)

The rule book for the district junior show in my area says lamb's bellies may be clipped or shaven, but wool should be approximately ⅝ inch long "all over the lamb." Let's see you do that!

A clean, well-fitted lamb ready to go into the show ring.

We could argue that the market wants a number one pelt, but we also know the market doesn't hand out ribbons. A number one pelt is currently worth one dollar more than a number two, and three dollars more than a number three.

I lean toward the philosophy that the kids operate in a different market. They learn a lot from their experiences, and they must be doing OK, because they make a lot more money with their sheep than the rest of us do.

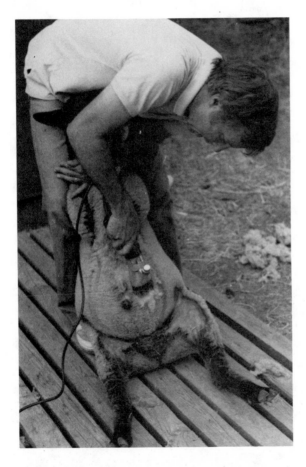

It is a common practice to shear the belly on market lambs about a week to 10 days before the show.

PLANS FOR WOODEN BLOCKING STAND

CUTTING LIST

Item	No.	Description
A	2	1x6 x 3'-6"
B	2	1x6 x 24"
C	3	1x8 x 3'-6"
D	4	1x4 x 17"
E	8	1x4 x 22-1/2"
F	1	1x6 x 22-1/2"
G	1	1/8" x 1/2" x 12" strap
H	8	1/8" x 1" x 6" strap
I	4	2-1/2" butt hinges
J	4	4" T-hinges
K	1	heavy screen door handle
L	1	1x6 x 33"

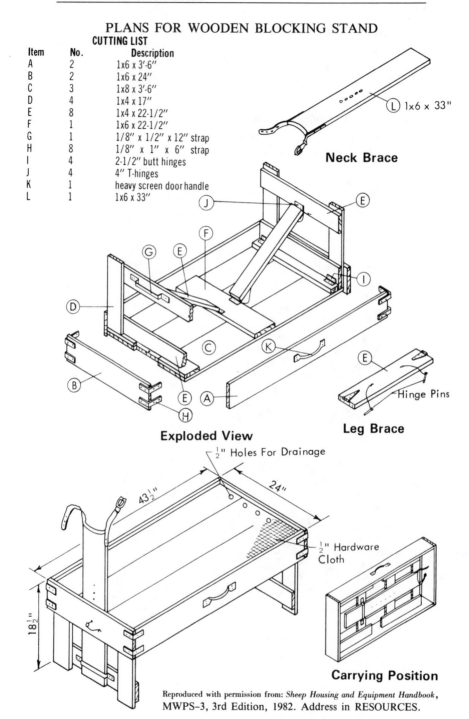

(L) 1x6 x 33"

Neck Brace

Exploded View

Hinge Pins

Leg Brace

½" Holes For Drainage

43½"

24"

½" Hardware Cloth

18½"

Carrying Position

Reproduced with permission from: *Sheep Housing and Equipment Handbook*, MWPS-3, 3rd Edition, 1982. Address in RESOURCES.

PLANS FOR METAL BLOCKING STAND

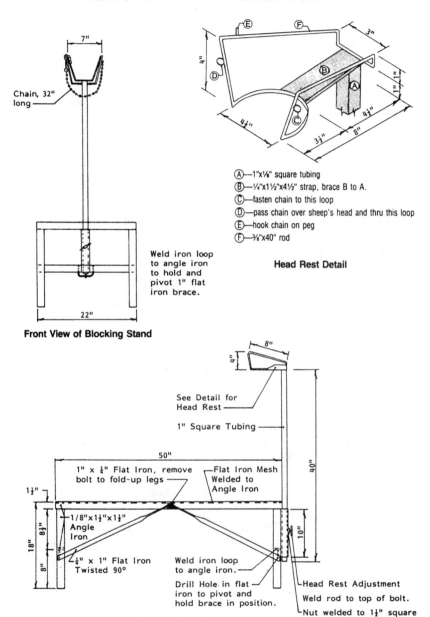

A—1"x⅛" square tubing
B—¼"x1½"x4½" strap, brace B to A.
C—fasten chain to this loop
D—pass chain over sheep's head and thru this loop
E—hook chain on peg
F—⅜"x40" rod

Head Rest Detail

Chain, 32" long

Weld iron loop to angle iron to hold and pivot 1" flat iron brace.

Front View of Blocking Stand

See Detail for Head Rest

1" Square Tubing

1" x ¼" Flat Iron, remove bolt to fold-up legs

Flat Iron Mesh Welded to Angle Iron

1/8"x1½"x1½" Angle Iron

¼" x 1" Flat Iron Twisted 90°

Weld iron loop to angle iron.

Drill Hole in flat iron to pivot and hold brace in position.

Head Rest Adjustment
Weld rod to top of bolt.
Nut welded to 1½" square tubing.

Side View

Reproduced with permission from: *Sheep Housing and Equipment Handbook*, MWPS-3, 3rd Edition, 1982. Address in RESOURCES.

Showing Market Lambs

Preparing a lamb for show is more than carding, clipping, and washing. There are a number of other essential tasks, including having the animal tame enough to handle and in proper show condition.

Now that you have the needed equipment and a good idea of how to begin fitting your lamb, let's talk about training and final preparations for showing. This may take some equipment too, but it doesn't need to be elaborate.

Nearly all of the kids with lambs use halters. These can be purchased, but they can also be made very cheaply from about six feet of nylon rope. The first six feet of rope from Dad's boat anchor is about right.

There is an old bulletin in county extension offices of many states which shows how to make a rope halter for a cow or steer. Halters for sheep or lambs can be made by reducing the dimensions. The information was originally printed in a Cornell University 4–H bulletin and is probably available in many forms.

A halter is almost essential for working with junior show lambs at home, as well as around the fairgrounds. You want to train the lamb so that the halter won't be necessary for showing, but it will be helpful in getting the animal under control.

A trimming stand is very helpful for taming a lamb as well as for fitting. Once you get the animal up on the stand, it's pretty much immobilized and will become used to being handled and worked with. They do, of course, jump off a few times; so you shouldn't go too far away when the lamb is on the stand.

My daughter once had a lamb jump off the stand, run around the barn, and jump into his pen — with the stand still attached to his neck! This wasn't a very heavy stand, but I would admit the lamb was a bit excited.

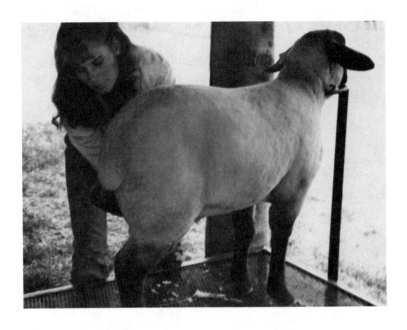

Working with the lamb on the trimming stand helps gentle the lamb and makes it easier to handle during the show.

If you don't have a trimming stand, the lamb can be tied to a wooden fence or post for washing, carding, and trimming; but a stand makes it a lot easier. It helps to pick the lamb's feet up repeatedly while it is on the stand or tied securely. This helps to tame the animal and accustoms the lamb to having its feet moved. This is important for later when you try to move the lamb's feet in the show ring.

The best time to begin training the lamb is a matter of opinion and depends upon the size and experience of the junior showman. A lamb that is not gentle enough for the kids to handle is not going to work, but one that is a total pet will probably go to sleep in the ring. For the smaller kids I would rather have the pet than the runaway.

Experienced showmen will say that you want the lamb nervous enough so that it will tense-up when the judge handles it. This makes the lamb feel harder.

I had to laugh at the comment of a sheep breeder a few years ago when she said, "The best way to make a lamb feel hard to the judge is to take him out of the pasture the day before the show and put him into the ring. If he tries to get away, grab him by the throat until he can't breath. A lamb that isn't breathing isn't going anywhere." I think she was exaggerating, but I can never be quite sure.

Most lambs will tame down if worked with for about three weeks before the show. Activities such as washing, trimming, and carding will help a lot in this regard.

The most frequent question asked at our fair is, "Would it be OK for the little kids to show the lamb with a halter?" This is debatable, I think.

Everyone who has been around sheep shows knows we don't show lambs with halters. We show cattle with halters, but we hold lambs by the throat. Why is that? Because I said so, that's why! (These kids come up with the darnedest questions.)

Leading with a halter is a good way to exercise the lamb and the easiest way to move it from one place to another.

Of course, the kids should have the animal trained so that they can handle it without a halter; and they shouldn't get upset if their lamb gets away in the ring. However, that's easy for me to say.

We should remember that we used to have 70 pound kids and 90 pound lambs in junior shows. Now we have 115 to 130 pound lambs, and the smaller kids are still about 70 pounds.

I tell the kids who ask me, "If you use a halter, you won't win the showmanship class. If your lamb runs off and you go back and do a good job of showing, you could win the class. How do you and your lamb feel about this competition thing?"

One thing that helps with a lamb that is difficult to control is to walk it around and tire it out a little before going into the ring. This may result in a lamb that is too relaxed to show at its best; but if you're worried about holding on, maybe a relaxed lamb is OK.

It has also been traditional to set the lamb by picking up the feet and placing them back down in the proper position. Again, we have a problem for the smaller kids with big lambs. While the kid is at the back end reaching for a rear hoof, the front end is heading for the barn. If the lamb has been worked with, it will set up fairly well when pushed one way or the other from the front.

If the showman will put a knee in the lamb's chest and throw it off balance or pull the lamb forward a few inches, the lamb will move a rear foot that is badly out of position and very often stand the way we want it. The bigger showmen can reach back and pick up the feet, but the little people can do OK if they'll push the animal around a little.

Once you have the lamb at the show, it's best to maintain the same ration you were feeding at home, except you will want to cut back on the amounts. Some showmen will hold all feed and water for a period of time before show. This depends upon the individual animal and the amount of fill needed for it to look its best. It is important not to have the animal filled up at the time of the show. A light feeding and control of water consumption the day of the show is generally advisable.

97

Most 4–H and FFA sheep project manuals do a good job of explaining what to do in the show ring. A little practice at home with someone acting as the judge is the best preparation for the skills needed in the ring.

Having a well-trained lamb gives the showman confidence and permits watching the judge.

Wool Care

I can well remember a cartoon done by western cartoonist Ace Reed back in the early 1970's. Old Slim opens the mailbox and says, "Hot dog! We got the check fer our wool! Now we can go to the cattleman's convention."

How true it is. The wool check always comes in handy, but nobody gives it any respect.

A few people around the country are trying to change all that and are hoping to educate sheep producers on the proper methods of harvesting and handling wool to produce a high quality product.

This is a worthy cause. Educating sheep producers is something a person could devote his whole life to. However, these people are willing to try, and they recognize there are a lot of small flock owners, like myself, who obviously have a lot to learn.

One of these educators is Ladd Mitchell, County Extension Agent, at Ephrata, Washington. In an interview with the *National Woolgrower* magazine a few years ago Ladd says, "The wool grown in the United States is as high of quality, or higher quality, than anywhere in the world. The New Zealand (shearing) instructors have told me this. Jim Walters of Pendleton Woolen Mills, who buys wool from all over the world, has said the wool here is as good as anywhere in the world, but the problem is that we don't 'harvest' it."

Wool has been taken for granted in the U.S. It has been seen as a by-product of lamb production and handled poorly as a result. This problem has often been blamed on the low price of wool in recent years; but I would have to admit, we didn't do any better when the price was higher.

So what can be done by the junior livestock producer or small flock owner to improve the quality of the wool marketed?

For many of us, we need to find a good shearer. My sheep have begged me to do this for years. When I shear them myself, the fleeces don't need to be tied; the blood holds them together.

The term 'half-blood fleece'' takes on a whole new meaning.

They tell me the fleece should come off in one piece, without second cuts that shorten and damage the wool. I suppose we owe it to the sheep (and the wool) to hire a good shearer.

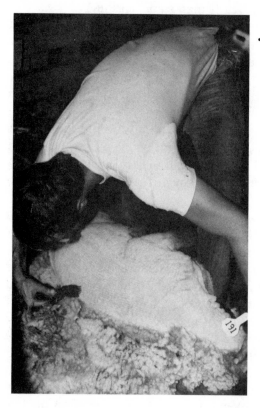

A good shearer is important to producing a quality wool clip.

There are not many good shearers in some areas, and it is often necessary to pool flocks in some manner to permit hiring a shearer from some distance. Many wool pools sponsor shearing days for this purpose.

I know it has been easier for some of us to do it the other way, but we should recognize what sort of product we are trying to sell.

No plastic twine should be used to tie fleeces, wool bags, or anything near the shearing floor. Many sheep producers are making it a policy not to use plastic twine anywhere on the farm because of wool contamination from this material.

Wool should never be stuffed into plastic feed sacks for the same reason. Fleeces should be tied with paper twine made for this purpose, and fleeces should be packed in wool bags. This is another reason for pooling small flocks for shearing — to have enough fleeces to fill a wool bag.

Sheep must be dry when sheared, and shearing should be done on a clean, dry surface. Fleece wool should be kept clean and bagged separately from tags or lamb's wool.

Separating lamb's wool, tags, and fleece wool retains the value of the wool clip.

If both black-faced breeds and white-faced breeds are being sheared, the white-faces are sheared first to avoid contamination of these fleeces with black fibers from the other breeds.

The value of wool from white-faced breeds can be increased by skirting the fleeces and separating belly wool from fleece wool. Small flock owners with a few white-face fleeces may be able to market these to home spinners for a much better price than can be expected on the commercial market. I should mention that not all white-faced sheep are created equal. Wool from some breeds is more desirable and will bring a higher price for particular uses.

Wool should be bagged and stored in a dry place until sold. It's best to put wool sacks on boards a few inches off the floor, to permit air to circulate beneath them. Wool will draw moisture from concrete floors and rot.

There's a lot more to know about wool and wool care than what I have told you here. Most county extension offices can furnish information on proper care and handling of wool. Many of us can pick up some new ideas from this information.

Of course, the wool incentive is a major part of the income to be derived from the wool crop in today's market. However, the person who sells for a higher price receives a higher total income from his wool.

SWINE

Selecting Market Pig Projects

The market pig is one of the simplest of junior livestock projects. Success is largely determined by two ingredients: a good pig and good feed. If you can locate these two items, you are well on the way.

However, before we are overcome with confidence, it might be well to remember that a good pig is not always easy to find; and sometimes only the pig can tell if we have the good feed.

By the time the pig can help us evaluate the feed, the project is nearly over; and we finish them off with everything from chocolate cake to the Cambridge diet, hoping to make market weight. I know there's no excuse for this kind of thing, but nearly everyone finds himself in this situation sooner or later.

Let's go back to the first critical step — the good pig. This is hereditary. The pig is born with the potential for good conformation and fast growth, or he is born with nothing but a curly tail and a nice personality. Even pigs from the same litter are sometimes quite different, if the genetics behind them aren't uniform.

The biggest difficulty in selecting market pigs for junior livestock shows is finding a breeder with some good pigs and the willingness to sell one or two. Once a good source of pigs is located, the rest generally comes easy. Even the question of what size pig you will need for your show date can usually be answered by the breeder.

At this stage the best predictor of these pig's performance is the breeding behind them.

Predicting how much a pig will weigh several months from now is always a risky venture, but the experienced breeder has a good idea of what his pigs will do.

Starting the project with a pig that is either too big or too small to reach proper weight for the show puts the junior livestock member at a serious disadvantage. Maintaining pigs on a reduced ration for long periods or attempting to bulk them up in the last few weeks almost always results in a bad experience for the kids, as well as the pigs.

Last year one of our 4–H club member's parents had the opportunity to buy some project pigs which a lot of people would have considered too small for our fair. The parent asked me what I thought.

I should have told him those pigs would do fine. However, I wasn't planning on moving away anytime soon, and my confidence in the pigs wasn't as high as it could have been.

So, I gave him the whole story: "If the pigs do well they'll make it fine, but sometimes they don't do so good, and a lot of people would suggest the kids buy bigger pigs, but if they are good pigs and they have good feed and plenty of water and they do well they'll be fine, but on the other hand ... "

To make a long story short, three pigs out of four did fine, but the other one did lousy. They were all from the same litter, too. I was glad they didn't just buy the one that did poorly.

That is one lesson in buying project pigs. There can be considerable variability in the performance of individual pigs, unless they are from very uniform breeding.

There are some guidelines to projecting how quickly a pig will reach market weight. The Ohio 4-H pig project manual estimates it takes about 100 days to get a 60 pound pig to a market weight of 200 pounds or about 120 days to feed a 40 pound pig to this weight. These numbers suggest a pig for a May 1 show should weigh about 60 pounds on January 20.

We should recognize that good pigs will do better than this, and that we'd like them to be more like 220 to 240 pounds in most shows. This is a reasonable place to start, however. We should make some allowance for poorer gains during hot weather or for less than ideal feeding environments. It's also possible to begin with a slightly older pig and hold him back a little if necessary. Some showmen prefer to do this, but it's important not to stretch this practice too far. A pig that hasn't been gaining well often doesn't show well, and a very hungry pig is always uncooperative and hard to handle.

What do you look for in selecting project pigs? Of course, you look for length and evidence of good muscling; and you look for one that looks healthy, which is another subject in itself.

I think the most important thing is to find a pig with good breeding. Not a registered pig probably, but a pig from stock that has the ability to grow well and finish as a good market hog. You are dependent on the breeder to have the kind of animal you want.

Several months ago a 4-H parent asked at a meeting, "What gives a pig muscle?" I said "heredity." Another club leader said, "exercise." Both answers may be correct to an extent. However, I maintain you can exercise them all you want; but if they aren't born with it, they are never going to have it.

It takes a good pig at a desirable show weight to go home a winner.

Project members and parents should remember to be considerate of the breeder when purchasing pigs. Spread of disease is always a concern for swine breeders, and they have to be careful about farm-to-farm traffic.

Showing pigs to prospective buyers takes time, and junior showmen should make their pig shopping visits as convenient as possible for the breeder. Some sellers of project pigs prefer to schedule a selection day when those who want pigs can come to the farm and choose their project animals. Many 4–H clubs or FFA chapters buy a group of pigs from one or two breeders and take the pigs home, where the members can then make a selection.

In some areas all of the project pigs are brought to the county fairgrounds, and all pig project members make their selection there.

How much should you pay for a project pig? I think something above market price for feeder pigs is to be expected. The breeder is providing a quality animal and often giving a choice of his best stock. He also deserves a premium for the extra effort involved.

On the other hand, I don't like to see junior showmen pay double the market price for project animals. I don't believe anyone's animals are that superior; I think we teach the wrong economic lessons when kids are encouraged to pay high initial prices in a gamble to have the grand champion.

There are plenty of good animals at a reasonable price, with an excellent chance the judge will like them just as well.

A good health program for pigs includes worming them early in the feeding period and possibly a second time in the middle of the period. Recommendations will vary by areas and management factors, but worming is essential.

The need for vaccinations or treatment for external parasites, such as lice or mites, will also vary by areas of the country. Your veterinarian or county extension agent are the best sources of advice on these matters.

Feeding Market Pigs

Once you have located a good pig, a proper feeding program will almost assure a good show pig at the desired market weight at the time of your show.

One of the first concerns in feeding and management of a pig project is worming of the pigs. This should be done soon after the pigs are obtained from the breeder, and a second worming 30 to 60 days later may be advisable in some instances. Your veterinarian is a good source of advice on the best time for worming and the most effective drug to use.

I would also suggest consulting your veterinarian about vaccinations for your market pig. Disease problems are quite variable by locality, and the local vet will know which diseases are prevalent in the area. Purchasing pigs from a breeder with a minimum of disease problems can also help in this regard.

Now, let's tackle the question of feed for a pig. There are two common methods for obtaining a pig ration: buy a complete prepared ration, or mix home grown (or purchased) grains with a prepared protein and mineral supplement. The choice between these options depends upon the availability and price of good quality grain and prepared supplements versus the cost and quality of the prepared feed available.

You can obtain good results from either system, if you have quality ingredients in the home-mixed ration or if you buy a good quality prepared feed when purchasing a complete ration. However, there are two major "ifs" in the previous sentence that must be considered.

110

Several years ago a leading swine nutritionist pointed out that you can't judge a pig feed by the feed tag alone and that the cheapest feed can be the most expensive in terms of cost versus results.

The nutritionist explained that you may buy a feed with 14 percent crude protein, but whether the feed contains the amino acids needed by the pig depends upon the sources of protein in the feed. The total protein level doesn't tell the whole story. If one or more amino acids are deficient, the cheapest feed may be the most expensive.

Therefore, you should be sure the feed tag lists at least an adequate level of protein, but you must also depend upon personal experience and the experience of others to determine the performance you can expect from a particular feed.

A magazine story also quoted the previously mentioned nutritionist as saying, "The cheapest food is the most expensive, always." I would suggest this could not have been meant literally.

A combination show feeder and tack box that can be made in a vo-ag shop. This gets the tack out of the aisles and makes a neater exhibit.

Many of us have certainly paid high prices for cheap feed. Let's say that the cheapest feed is *sometimes* the most expensive. Performance and price are both important to a junior pig project member.

On farms where home grown feeds and good mixing facilities are available, a ration which utilizes these feeds is usually the most economical. Where corn, barley, or wheat are available, these grains often form the base for home-mixed rations.

A few cautions are in order for those who mix rations at home. One is to be sure ingredients are well mixed and that supplements containing antibiotics or other additives are mixed and used properly.

To illustrate that care is in order, I'll relate an incident brought to my attention several years ago. In this instance the feed store was selling a brand of protein supplement in 50 pound bags. The supplement was to be mixed with grain at ratios that would put one bag of supplement with several hundred pounds of grain for swine rations. In a very similar bag, the store also carried a supplement with a high level of medication to combat a particular swine disease. This was to be mixed at a ratio of one bag per ton of feed.

At least a few parents of 4-H members bought the wrong supplement mixture and mixed it as they had previous batches. Several pigs consumed a lot of medication before the owners learned of the mistaken purchase.

Caution is also in order when purchasing complete rations. Some feeds contain low levels of medication for younger pigs and require a withdrawal period before slaughter. It is important to read feed tags carefully.

It is generally recommended that pigs under 50 pounds receive a ration containing about 18 percent crude protein; pigs 50 to 100 pounds receive a ration with 16 percent crude protein; pigs 100 to 160 pounds receive 14 percent protein; and pigs over 160 pounds receive a 12 to 14 percent protein ration.

These are averages that must be taken with a grain of salt, remembering what was just said about the hazards of looking at total protein percentages. A ration with more than the recommended protein level won't hurt the pigs, but generally costs more. This is why we reduce the protein level as pigs become larger and their protein need changes.

Should you hand feed the pigs or put them on a self-feeder? I think this is more of a concern to the people than to the pigs. Pigs will do fine on a self-feeder, if it is built to prevent wastage. Most showmen prefer to hand feed to control intake the last two weeks before show. This gives an opportunity to trim up the pig's belly by reducing his consumption. (Livestock shows cause us to do funny things.) A pig that is becoming too large or overfat well before show can also be held back with controlled consumption.

Many 4-H leaders, FFA advisors, and parents like to see the kids hand feed — with good reason sometimes. Hand feeding ensures a regular check on the animals for one thing. It's embarrassing to be unable to report which day the pigs ran off.

A survey would show parents like to see the kids feed pigs twice a day, kids prefer the self-feeder method, and the pigs don't really care as long as there is no confusion and no slip-ups.

A self-feeder is one good way to feed market pigs.

The Washington 4-H pig project book gives the following figures for the amount of feed consumed and estimated daily gain for pigs at various weights.

Weight of Pig (pounds)	Weight of Feed Eaten Each Day (pounds)	Daily Gain (pounds)
25	2.0	0.8
50	3.2	1.2
100	5.3	1.6
150	6.8	1.8
200	7.5	1.8

The pig feeding facility should include a system of providing fresh water at least twice a day and for keeping water available at all times. A self-serve water system giving access to cool water is best, especially during the summer months. Shade is also important.

The automatic watering device needs to be securely fastened, or the pigs will take it apart and build themselves a swimming pool.

The best way to tell whether a pig is gaining well or approaching show weight is to weigh the animal on a good scales. In reality, however, most of us don't have livestock scales, and loading a pig in the station wagon becomes irksome if you have to do it often. It's possible to get a fairly good estimate of a pig's weight by measuring the heart girth and length of the pig with a commercial weigh tape available at many feed stores.

If you don't have the weigh tape, I have found the following formula will provide a good weight estimate for pigs between 100 and 250 pounds. Simply measure the pig's heart girth (chest measurement) in inches just behind the front legs. Multiply this number times itself (square it), and divide by 8.5.

$$\text{Weight of pig} = \frac{\text{Heart girth}^2}{8.5}$$

Example: Pig's heart girth measures 40 inches.

40 × 40 = 1600 divided by 8.5 = 188 lb. (pig's weight)

This is not as good as scales, but will usually estimate to within 10 or 15 pounds. It's best to check the formula with scales occasionally to determine accuracy of the estimate.

Fitting and Showing Pigs

You have a good pig, and you've fed him to the proper show weight, but there's more. The pig has to be properly fitted and shown to do well in tough competition.

A market pig is relatively easy to prepare for show, but a good fitting job and a skilled showman can certainly help a good pig show even better. It's no accident that the top animal in a tough class is almost always clean and well-fitted in addition to having superior conformation.

Sometimes junior showmen may get the idea that animals don't need to be clean and handled properly in market classes, as long as they're ready for the showmanship class. This is a big mistake. I am amazed to see kids come into market classes with a dirty, unbrushed animal and then come back the next day with the same pig all dolled-up for showmanship.

Showmanship is not meant to be a different game from market classes. The same preparation and skills that are expected in showmanship will also help the animal place better in a market class. If we were to bring two pigs of similar conformation into a market swine class but leave one of them dirty, I'll guarantee the clean one will win every time. The clean, well-fitted animal always looks better.

There may be shows that have found ways for judging showmanship and market classes together to help correct the kids' misconception that showmanship is different. It should be possible to judge the two classes together, but I have never seen it done.

This young showman has the pig between him and the judge
and is doing a good job of watching the judge.

Preparation for the show begins weeks before show day
with exercising and training the pig. The ideal situation for
training is a large pen or alleyway, but most junior showmen
have to do their best with whatever facilities are available.
Early training should concentrate on getting the pigs used to
the cane or whip and acquainted with the showman.

The pig should be tame enough that the showman can put
his hands on it, but we don't want to make a pet of the pig.
We want to avoid undue excitement or anything that teaches
the animals bad habits. It's best not to fight with them.

If pigs have not been in a clean well-bedded pen, it may be
necessary to wash them a couple of weeks before the show to
get caked mud or dirt off such places as the knees. Most pigs
will only need a couple of washings, a week or so before going
to the show, if they are kept in a clean pen after washing.

Daily brushing, beginning a couple of weeks before the
show, will improve the appearance of the hair coat and help
clean the pig. Hair on the ears should be clipped both outside
and inside the ear two or three days before the show. This

117

should be done carefully with scissors or small hair clippers, such as barber clippers.

If electric clippers are used, many pigs will bite the electric cord at the first opportunity, creating a dangerous situation. Some people protect the cord by inserting it through a split rubber hose. Hand clippers are much safer but hard to find. I believe the kids can do fine with a pair of scissors.

Hair is normally clipped from the upper third of the tail or all the way down to the switch, depending upon what part of the country you are in. The underline can be trimmed starting just above the teat line and trimming down and under the belly.

1

1 Many showmen use electric hair clippers for clipping ears and tails. Make sure the electric cord is out of reach of other pigs as well as the one you are trimming.

2 The pig's ears are clipped outside and inside. It isn't necessary to clip hair deeply within the ear.

3 The tail is generally trimmed from the base of the tail to the brush. This may vary in different parts of the country.

2

3

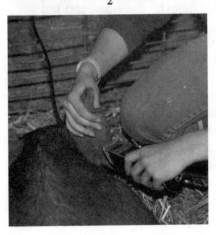

In some shows the entire underline isn't trimmed; only long hairs on the belly are clipped. Long hairs around the nose, eyes, and mouth should also be clipped.

It is not usually necessary to trim feet on market hogs, unless there is excessive growth preventing the pig from standing and walking correctly. If feet are trimmed, this should be done at least two weeks before the show to allow time for healing of any tenderness or mistakes in hoof trimming.

Health papers are required for entry to shows in some areas of the country and should be obtained well ahead of time to avoid last minute hassles. Withdrawal times must be observed for any medications used in feed or shots given to the pigs to be sure particular medications are not used too close to slaughter.

When pigs are to be loaded for hauling to a show or moved to places they don't want to go, all types of fights and wrestling matches should be avoided whenever possible. Many showmen have found that a five gallon bucket placed over a pig's head turns a squealing dervish into a confused and placid porker. You can then follow him anywhere, because the pig will only travel in reverse, if he can't see out of the bucket.

All you have to do is aim his tail the preferred direction and keep up with a pig that is staggering backwards. This is much better than the squealing, beating, and hollering often seen in the hog barn. (A smaller bucket and faster footwork is required for smaller pigs.)

Pigs may be washed the day before the show or several hours before show time. A good brushing after washing and again before showing will help the pig's appearance. A light spray of water from a spray bottle and a good brushing just before show will also help. Be sure the pig is dry before going into the ring, however.

We used to apply all sorts of baby oils and powders to pigs to slick them up, but this is no longer permitted in many shows. Meat packers have some severe problems with slick pigs. Most youth shows have either outlawed oils and powders or discourage them severely.

The day of the show, many showmen recommend the pig be fed lightly — about half of a normal feeding. This feeding should be two hours or more before showtime. Water should also be restricted. We don't want the pig too full, but you don't want him so hungry that you can't get his attention in the ring. A pig that is very hungry is also more likely to start a fight.

Project materials for 4–H and FFA members contain good suggestions for showing pigs while in the ring. Customs vary in different parts of the country. When I moved from Ohio to the Northwest, I was surprised to see the kids in the West hold onto the crook end of the cane when showing pigs. In the Midwest they used to show with the crook of the cane next to the pig. Now most junior showmen in the Midwest show with a small whip rather than a cane. Either way, the principles of showing pigs are the same.

It helps to have a few practice sessions at home with someone playing the part of the judge. Learning how the pig responds and how to move and stop the animal can only be achieved through practice.

A well-fitted pig handled by a good showman.

Breeding Swine Projects

Breeding sows may not be as popular for small and part-time farms as they once were, but there is still a place for the old "mortgage lifter" as hogs were once called. The opportunity to raise feeder pigs exists in areas where pigs are hard to find, as well as in those regions boasting a pig behind every corn crib.

Producing feeder pigs for sale to junior livestock showmen can be especially profitable in areas such as the Northwest where good quality pigs aren't always easy to find. In these situations, a few sows (maybe only three to five) can turn a profit even when market hog prices are low.

Elaborate facilities and expensive feeds aren't needed for raising a few hogs. Nor is it necessary to spend a lot of money for purebred breeding stock. Crossbred sows are generally more prolific and better mothers than purebreds and are probably a better choice for most beginning hog raisers.

Purebred boars, on the other hand, are often recommended to assure more uniformity of the pigs produced. While it may seem questionable whether it is economical to own a boar to service just a few sows, the value of disease prevention and predictable quality of the offspring will usually justify the added expense of owning the boar.

When selecting a breed (or breeds) it is important to consider which breeds are popular and available in your area. The person with only a few sows can't afford to travel all over the country to obtain replacement stock. White-skinned hogs are preferred by packers in some regions, but there are also advantages to the colored breeds. White breeds crossed with colored breeds will often produce white pigs with only a few dark spots.

Gilts should be structurally sound with good feet and legs and should have a minimum of twelve functional, well-spaced nipples. A moderate amount of muscle is desirable, but excessive muscling should be avoided. Extremely heavy-muscled sows are seldom good mothers.

Clean, dry quarters are important for pig health.

Good crossbred gilts can often be purchased from commercial hog producers as feeder pigs for a relatively low cost. Another source may be the higher placing gilts in market classes at the local fair, although market sale prices or local sale rules often eliminate this option. State and local swine breeder's associations schedule sales throughout the year where purebred breeding stock is sold. Check with your local county agent for sale dates or for other local sources.

Boars can sometimes be purchased from purebred breeders at weaning or shortly after for very affordable prices. Service age boars can be obtained at state or breed association sales.

Production records are valuable for making selection decisions. Both boars and gilts should be selected from large litters of ten or more pigs and from sows with good production records over a period of years.

Both boars and gilts reach sexual maturity at approximately eight months of age. Breeding them earlier than this age usually results in small litters. Gilts should farrow their first litter at one year of age.

Breeding programs for a few sows are easily planned and managed. Good show pigs are usually close to six months of age at show time. The gestation period for hogs is three months, three weeks, three days, (and three o'clock in the morning). Once we have identified the fair or show dates for which the pigs are intended, we can back off six months for a desirable birth date, subtract about four months for the gestation period, and will decide to breed the sow 10 months before the pigs' expected show date. Sows producing pigs for an early August show should be bred approximately October 1 to farrow in early February.

Sows should farrow twice a year to be profitable. This means one of her two litters each year will likely be sold as feeders or fed for market rather than going to a junior show.

Sows will generally be in standing heat three to four days after their pigs are weaned, regardless of the age of the pigs. Pigs are usually weaned at six to eight weeks. The sow can be bred at this time or held one or two heat cycles if being bred to farrow for a particular show date. Heat cycles average 21 days.

Facilities need not be elaborate or expensive. The sow's greatest needs are shade and fresh water in the summer and draft-free shelter in the winter.

Single unit 'A' frame houses faced away from prevailing winds provide adequate shelter for adult hogs during the winter. Similar units with doors on both ends and built-in stalls work well for farrowing units. Heat lamps can be used in the houses at farrowing time, but all safety precautions should be observed. Sows and heat lamps can be a dangerous combination if proper guarding of the lamp is not provided.

Mud holes and wallows aren't necessary or advised. Mud holes only serve to produce flies, disease, and odors. Pens should be dry and fitted with a water source which can't be rooted around or destroyed.

Welded wire panels and steel posts make excellent movable pens around farrowing houses. Hogs are very sensitive to electric fences and these work well for confining adult hogs to lots or pastures. Tighter pens are needed for small pigs.

A good health program is important to profitability with swine. Sows should be wormed before each farrowing, and boars should be wormed at least once a year. Piglets should be treated for worms at weaning.

New piglets should be given iron at one to two days of age. Both injectable and oral iron sources are available.

Lice can be controlled in adult hogs with pour-on systemic insecticides. If these are applied to nursing sows, lice will be controlled on the nursing pigs as well.

Boar pigs not being kept for breeding should be castrated at less than two weeks of age.

Following a few simple rules will go a long way toward maintaining a disease-free herd. These rules include

1. Buy healthy, disease-free breeding stock.
2. Isolate new hogs for at least 21 days after purchase.
3. Don't let "outside hogs" onto the farm.
4. Be careful of visitors, and don't visit other hog operations in shoes and clothing worn around hog facilities at home.
5. Move hog lots occasionally.
6. Don't let mud holes develop.
7. If using farrowing houses, clean and move them between farrowings. Don't farrow on the same site more than once in two years.
8. Keep facilities and equipment clean.
9. Maintain a good nutrition program.
10. Watch animals daily and treat immediately if problems appear.

Feeding Breeding Sows

The first principle in feeding breeding sows is to consider the stage of production in designing rations. The feed requirements for a lactating sow may be four to five times her requirements during early gestation. Rations must be adjusted throughout the year to maintain health and productivity in the sow, as well as solvency in the producer's bank account.

Feed is always the major production cost when raising hogs. On a per-head basis, feed expense is even greater for a few animals than for larger numbers because quantity discounts are not available. The lack of grinding and mixing facilities also works against the small producer and increases feed costs.

There are ways for the junior livestock member or small producer to hold feed costs to an economical level, however, if we understand the hog's nutritional requirements and make the best use of the feeds available. If we have to buy commercial sow feed, one bag at a time, it's very hard to make money at average hog prices.

There are often inexpensive sources of grain available to kids with the right contacts. Cleaning out farm storage bins or elevator legs can provide cheap grain for a little work. Screenings from seed cleaning operations are another opportunity in some areas. (Be sure of what's in the screenings, and never feed treated seed.)

Cleaning railroad box cars or cleaning up spills around the tracks is also a source of hog feed. Caution is in order when scrounging feed of any sort to make sure of all ingredients and not use something like treated seed.

Balancing a ration for hogs is more difficult than for ruminant animals, such as cattle and sheep. The hog's digestive system is a simple stomach and much like the human digestive system. Because swine have no digestive vat, such as the rumen in cattle, there is no bacterial breakdown of feeds before the gastric digestion in the true stomach. This means hogs are unable to digest large quanities of roughage, as the ruminants do.

Hogs also need complete proteins with the proper balance of amino acids, whereas the ruminant animals can reassemble amino acids and manufacture proteins through bacterial action in the rumen. Hogs require B vitamins in the diet, while ruminants get these from bacterial activity.

All of this means rations for breeding swine must be well balanced and provide amino acids and vitamins in the proper amounts. It also says that many feeds formulated for ruminants are not good for hogs and certain additives to cattle and sheep rations, such as urea, can be toxic to hogs.

One solution to the balanced ration question is to buy a complete feed which includes grains for energy and has all of the necessary amino acids, B vitamins, and minerals included. This may be best in some situations. However, a more economical course if good quality grains are available is to feed a commercial protein supplement and mix it with grains or other feeds that can be found at a good price. These supplements give directions on the tag for mixing and feeding to various ages and classes of animals.

Grains commonly fed to hogs in many parts of the country include corn, barley, wheat, and oats. Rye can also be used successfully if fed at less than 50 percent of the ration. In the Northwest, wheat with rye contamination or low quality wheat are common feed grains and may be available from farmers at economical prices.

Grains are commonly rolled or ground for swine feeds, but some grains, such as wheat, may be fed without grinding under certain conditions. Grinding gives better feed efficiency with most grains, but the cost of grinding can exceed the benefit in situations where whole grains are cheap and grinding is inconvenient or expensive.

Breeding sows may need up to 20 pounds of feed per day during lactation to keep them milking and to put them in condition to breed back after the pigs are weaned. At this level of feeding there is no room for fillers. High fiber feeds, such as forages or cull fruits or vegetables, should be limited. Condition of the sow and the number of pigs suckling should help determine the amount fed.

Amino acid levels and B vitamins are important at this time as are the high energy grains. A commercial protein supplement should be fed with available grains to assure the proper balance of amino acids, B vitamins, and minerals.

The sow also needs a balanced diet during gestation, but her nutrient requirements are greatly reduced during this stage. Sows may need only about four pounds of grain ration per day during early gestation, and care must be taken to prevent sows from getting over-fat during this period.

Because four pounds of feed won't keep the sow satisfied, high fiber feeds are used as fillers. This is sometimes provided

Feed requirements for the sow are greatly increased during lactation.

by legume pastures or alfalfa hay. Higher fiber grains such as oats are also good gestation feeds.

It is often best to hand-feed gestating sows because of the smaller amount of feed needed. If mixing equipment is available or a ration is commercially mixed, enough filler can be added to permit self-feeding without getting the sows excessiveley fat. This may add to feed costs, however.

Boars can be maintained on rations similar to those of gestating sows. Slightly more feed per day may be required, depending upon the size of the boar and how much he is being used in service.

Things like cull fruits and vegetables can be fed, but such things as raw dry beans contain toxins and should not be fed to hogs. Raw potatoes are not digestible to hogs, but cooked potatoes are OK. Root crops such as beets or carrots are digestible and provide some vitamins but little energy because of their high water content.

Garbage is not a feasible feed source for small numbers of hogs because of the licensing and cooking requirements.

Extra cattle feed should not be fed to hogs because of the possibility of ingredients such as urea or other additives which are toxic to hogs. If the total ingredients of cattle or sheep feeds are not known, they shouldn't be fed to hogs.

Salt should also be included in swine rations and can be fed free choice as loose salt. Salt is not dangerous to hogs unless they are salt-starved and gain access to large amounts without having water available.

I was taught that a combination of salt and water in the same container is also hazardous to hogs. Like many things I was taught, I can't personally vouch for that; but I keep the salt container dry just in case.

Let's not forget to provide clean water. Automatic waterers are best and should be leak proof or on a base of some sort to prevent the sows from making a large swimming pool.

DAIRY
CATTLE

Selecting a Dairy Heifer Project

I can well remember my first FFA dairy heifer project. Her name was "Such and Such, Something or Other, Snowball." I remember the Snowball part because she was nearly all white, like a snowball. She was built like one, too — just as round and smooth as she could be.

I thought she looked pretty good (nice and round and fat), and the judge at the county fair liked her OK, too. He placed her second in a class of four. Looking back now I can see that the judge and I were both wrong, unless that happened to be a pretty lousy class of heifers.

That heifer just continued to get bigger and rounder until calving time, and then we could see there was a problem here. She was too round and had very little dairy character or femininity. Her udder finally developed into the approximate size and appearance of a volleyball; and she never did give over 40 pounds of milk the best day of her life, which wasn't all that long.

This wasn't exactly what I had in mind; but I kept her two years, bred her to the biggest bull in the stud book, and had some very nice visits with the vet each calving as we tried to pull the biggest bull calves you ever saw out of that ungrateful heifer. One of the calves lived, too, just to prevent a total loss from the enterprise.

So why am I giving advice on selecting dairy heifers? Well, it's like your dad says, "So you won't make the same mistakes I made." If you're going to make mistakes, at least they should be new ones.

Selecting a dairy calf, especially the first one, can be a real challenge to a person's judgment and foresight because most of them are chosen as young and immature animals. Attention to good type is both practical and essential in the selection of animals for showing or in the establishment of a foundation herd. The structurally correct animal will always catch the judge's eye, and animals of desirable type generally produce and reproduce for a longer period of time.

A yearling dairy heifer of desirable type

A beginner may ask what we mean by "type." Type is the general form, structure or character of a particular animal. Simply put, it is the way an animal is put together — its conformation and eye appeal.

When selecting animals for show, the showman needs to learn the qualifications for the breed and to select animals that, when grown out and fitted, will be attractive to the judge.

Except for differences in color, size, and head character, all dairy breeds are judged on the same standards, as outlined in the *Dairy Cow Unified Score Card*. Anyone showing dairy cattle should be familiar with the unified score card and the criteria it suggests for conformation.

It should be noted that the *Dairy Cow Unified Score Card* was changed a few years ago, with more emphasis now placed on the udder, feet, and legs. If you have an old card without the new changes, it would be best to get a new card.

Where do we begin on selection? First, you'll have to choose a breed, and that could be easy because you may already know which breed you like. Or you may have a particular breed that is readily available. The breed chosen is important, so consider all of the factors before you buy.

Youngsters shouldn't buy a certain breed just to be different or to avoid stiff competition in the show ring. They will learn more if there is strong competition in both numbers and quality of animals. A good dairyman will generally be successful with any breed.

The next question is whether to buy a purebred or grade animal. This decision will be greatly influenced by the ultimate objectives. Are you starting a foundation herd, or are you showing at breed association shows as well as youth shows? Remember, however, that registration papers don't assure that your calf will be the kind of cow you want. My heifer was registered, rest her soul.

Good calves can be found in most areas of the country, but there are places where a little travel may be required to locate the calf you want. If possible, buy from someone you know has a high producing herd that is free of diseases.

If the decision is to go the purebred route, and this will be your first experience at raising a calf, there may be limits on where you can buy an animal. Some breeders won't sell to just anybody because of the chance of damage to the herd's reputation if the animal is not raised properly.

If the calf is primarily for show purposes, it is important to buy one that was born at the right time. Dairy cattle are shown in age groups, and primary consideration should be given to calves born near the beginning of their age class. A calf near the top end of its age class often has a big advantage in size and appearance over younger calves in that class.

It is best to select from calves that are at least three to four months of age. There is less danger of death loss, and a more reliable evaluation of the calf's future can be made at this time than is possible at an earlier age. One way to determine if the calf is of desirable size for its age is to compare its heart girth measurement with the established normal measurement for that breed at a particular age. The chart *Timetable and Growth Standards for Dairy Calves and Heifers* accompanying this chapter gives expected heart girth measurements for calves of various ages.

Holstein heifers at breeding age. Size for age is an important factor in selection of breeding heifers.

Look for an alert heifer with good stature (size and scale) and length of body. She should be well-proportioned and have good dairy character which is evidenced by a clean-cut neck and sharpness over the withers. The rump should be almost level and have good width. Body capacity will be shown by a wide chest and well-sprung, open, and deep ribs. The rear legs should be straight as viewed from both the side and the rear.

If you don't have previous experience in selecting heifers, it would be well to take an experienced person with you to help make the selection.

Udders are very important to dairy cows. You have probably heard the statement, "No udder, no cow." This is very true in the dairy business. Since it isn't possible to tell much about udders on young calves, you may want to ask to see the calf's mother and milking daughters of the calf's sire.

Size, shape, and quality of the udder as well as attachments are all important. Probably the best indication of a strong attachment is a good crease down the center of the udder, with teats hanging straight down or pointing slightly inward. The quarters should be evenly balanced and symmetrical. A good quality udder is characterized by a soft texture, is pliable and elastic, and is well collapsed after milking.

One other detail that may be important to you is color pattern. If an animal is registerable by one of the dairy breed associations, no discrimination against color or color pattern will be made.

If you have a calf with white knees, you'll find that knee stains are the most difficult to get rid of. Not a big thing, really. I would hate to encourage selection based upon the color of knees.

Let's not forget the production records. After the heifer's show days are over (as a heifer, anyway) you want her to be a good producer for the herd and capable of having offspring with good production potential. Again, someone with dairy experience can be helpful in interpreting production records.

Selection will always contain a certain amount of guesswork, especially when buying young calves. However, a systematic approach combined with some knowledge and good judgment will reduce the guesswork and improve your chances for a good experience.

Heart Girth Measurements for Normal Growth of Dairy Calves and Heifers

Age Months	Holstein, Br. Swiss Inches	Guernsey Inches	Jersey Inches	Ayrshire Inches
Birth	30.0–32.0	28.0–30.0	24.5–26.5	28.5–30.5
1	33.5–35.5	30.5–32.5	28.5–30.5	31.0–33.0
2	37.0–39.0	33.5–35.5	31.5–33.5	34.5–36.5
3	40.0–42.0	37.0–39.0	35.5–37.5	37.5–39.5
4	43.5–45.5	40.0–42.0	37.0–39.0	41.5–43.5
5	47.0–49.0	43.0–45.0	40.5–42.5	44.5–46.5
6	50.0–52.0	46.0–48.0	43.5–45.5	47.0–49.0
7	52.5–54.5	48.5–50.5	46.0–48.0	50.0–52.0
8	53.5–55.5	50.5–52.5	48.5–50.5	52.0–54.0
9	56.0–58.0	52.5–54.5	50.5–52.5	54.0–56.0
10	57.5–59.5	54.0–56.0	52.0–54.0	56.0–58.0
11	59.5–61.5	55.5–57.5	54.0–56.0	57.0–59.0
12	61.5–63.5	57.0–59.0	55.5–57.5	58.0–60.0
13	62.0–64.0	58.0–60.0	56.5–58.5	59.5–61.5
14	63.0–65.0	59.5–61.5	57.5–59.5	61.0–63.0
15	64.0–66.0	60.5–62.5	58.0–60.0	62.0–64.0
16	65.0–67.0	61.5–63.5	58.5–60.5	63.0–65.0
17	66.0–68.0	62.5–64.5	59.5–61.5	64.0–66.0
18	67.5–69.5	64.0–66.0	60.5–62.5	65.0–67.0
19	68.0–70.0	64.5–66.5	61.5–63.5	65.5–67.5
20	69.5–71.5	65.0–67.0	62.0–64.0	66.5–68.5
21	70.5–72.5	66.5–68.5	63.0–65.0	67.5–69.5
22	71.5–73.5	67.0–69.0	64.0–66.0	68.0–70.0
23	72.0–74.0	68.0–70.0	65.0–67.0	69.0–71.0
24	72.5–74.5	69.0–71.0	65.5–67.5	70.0–72.0
27	73.5–75.5	69.5–71.5	66.0–68.0	70.5–72.5

Source: Extension Mimeo 2556 prepared by B.F. Kelso, Extension Dairy Scientist, Western Washington Research and Extension Center, Puyallup, WA.

Care and Management of the Dairy Heifer

The majority of 4-H and FFA members with dairy heifer projects are from dairy farms. They may have a ready supply of good quality dairy animals; and their parents know how to feed and care for heifers, or they wouldn't still be in the business.

This is not always the case, however. Many dairy projects begin with a purchased heifer from another farm, and some result from the offspring of the family milk cow. It's important for these calves to have a good start if they are to develop to their full potential.

Proper care for a dairy heifer need not be complicated, but a heifer is an important investment, requiring good care if she is to remain healthy and productive. If the heifer is from a large breed, such as Holstein, she should weigh about 1,200 pounds at first calving around 24 months of age.

This means she must average 1.5 pounds daily gain from birth to calving. Feeding, housing, and management are all important to attaining the goal of a healthy, productive heifer.

One of the first requirements after the calf is born, or after getting the calf from the breeder, is to positively identify the heifer and set up a system of records for important events, such as vaccinations, growth rate, health status, etc. Identification can be done by a sketch of the animal or by photographing the calf from both sides and the front.

A more permanent type of identification is the tattoo. If the calf is to be registered with a breed association, be sure to check the association rules before tattooing.

It's important for the heifer to be vaccinated against brucellosis at the proper age. At this time the veterinarian will place a brucellosis shield tattoo in her right ear.

The heifer should be dehorned by one or two months of age, especially if she will be shown at a fair. Most fairs won't permit showing of dairy animals with horns.

Removal of extra teats should be done as soon as they can be distinguished from the four main teats. It is well to be careful with this job, as there have been cases of the wrong teats being removed. A circular called *Removing Horns and Extra Teats From Dairy Calves* is available from county extension offices in some states.

If purchasing a heifer, rather than raising a calf born on your farm, it is generally recommended that she be at least three or four months of age when the selection is made. However, for purposes of this discussion, let's begin with care of a very young calf and work up to the older heifer you may select if buying off the farm.

First let's consider the needs of a newborn calf and then look at some feeding basics for a heifer up to six months of age. The newborn calf needs to receive colostrum milk soon after birth. Colostrum is the first milk produced by the cow after calving and is important to the calf because of the antibodies it contains. These antibodies help provide resistance to many of the diseases present in the calf's environment.

Antibodies can be absorbed through the calf's stomach in large amounts soon after birth, but within 24 hours this absorption ability has declined to near zero. Ideally, the calf should consume colostrum within the first hour after birth and continue to receive this milk for the first three days of age.

Colostrum can be frozen and saved for use with other calves, also. When thawing frozen colostrum, you should be careful not to overheat it, as this will destroy antibodies. Some booklets suggest thawing at room temperature; others say not to heat above 140° F. I would go with the latter. If you thaw colostrum at room temperature, the calf may be too old to use it by the time you get it thawed!

Some recently published information states thawing colostrum in a microwave is OK, also. A research veterinarian at the University of Idaho has found the microwave safe for thawing colostrum, if the oven is set at 60 percent power. He has found quality concentrations to be three times greater in colostrum thawed with the microwave, as compared to thawing by warm-water methods. He stirs the mixture as it thaws and cautions against setting the oven at more than 60 percent power.

When buying very young calves, the question of whether the calf received this early colostrum becomes important, also. This is one of the risks often taken when buying calves at a sales yard. We generally have no idea what care these calves had at birth. (Of course, the sales yard is not where you want to buy a dairy heifer project, anyway.)

My information says calves should receive from 8 to 16 pounds of colostrum milk per day for the first three days. This amount depends upon the size of the calf and should be split into two or more feedings per day.

I am nervous about giving absolute numbers for feeding recommendations. A Jersey calf may be half the size of the Holstein. We must use some judgment.

After the first three days, calves should receive whole milk or good quality commercial milk replacer equal to about 8 to 10 percent of their bodyweight per day. Colostrum milk is good for older calves, too, when it's available. It should be diluted with water at a ratio of about two parts water to one part colostrum when fed to older calves.

The experts say over-kindness in milk feeding kills more calves than any other particular error. All young calves will act hungry, and it's better to slightly underfeed the milk than to overfeed.

Dairy calves can be fed with a nipple bottle, a nipple pail, or can be taught to drink from an open pail. Regardless of the system used, all containers must be cleaned frequently.

The sooner a young calf begins eating dry feed the better. The calf should be provided with a small amount of high quality calf-starter (grain mixture) in a feed box within the first few days of age. The calf won't eat much dry feed for a few days, but the process can be speeded up by putting a small

139

Calf hutches are a popular management system in many parts of the country.

handful into the calf's mouth right after feeding the milk. Calf starters should contain 14 to 16 percent protein.

Good quality alfalfa hay should also be made available within the first week, as should fresh water and trace mineralized salt. The calf will soon be eating small amounts of calf-starter (grain mixture), and the amount should be increased until the calf is eating two to three pounds of this mixture at three months of age.

A calf doesn't need a fancy place to live, but she should have plenty of fresh air and a dry place to lie down. Yakima, Washington, Extension Agent Eddie Thomason has a novel suggestion for determining if calf pens are dry enough. Eddie says that if the owner of the calf would just get into the pen and sit down on the bedding for about five minutes, he will find out if it is dry or not. (This is a good technique for the youngster who always says, "Yes, it's dry.")

Calves should be kept in individual pens until they are weaned. This will help decrease the chance of spreading disease from one calf to another and also makes it possible to watch the calf's eating habits more closely. Penning individually also prevents the problem of calves nursing each other.

About a week after weaning, calves may be placed together in small groups. Of course, good sanitation practices must be observed whether calves are raised individually or in groups.

The calf should be provided with a calf-starter (grain mixture), good quality hay, water, and minerals at an early age.

The calf can be weaned from milk replacer at six weeks of age if it is doing well and eating at least one pound of grain mixture per day. At six months, the calf should be receiving about three to five pounds of grain mixture per day.

For those who would prefer to mix a grain ration at home, rather than buying a commercial calf-starter, the following is a recommended grain mixture for young calves.

Coarse ground or rolled barley or corn......50 lb.

Coarse ground or rolled oats.............40 lb.

Linseed or soybean meal.................10 lb.

Total.......100 lb.

One pound of trace mineralized salt and 0.1 pound of steamed bonemeal or dicalcium phosphate should be added to each 100 pounds of this concentrate.

By the time the heifer is six months old, she can handle a lot of hay and some pasture. Pasture grasses contain a lot of water and can cause a pot-belly, in addition to being inadequate in nutrition for heifers under one year of age. These heifers should receive hay and grain rations in addition to pasture.

A feeding program for dairy heifers should keep them in good condition and growing properly, but we don't want to get them fat. Research has shown that over-conditioning of young heifers lowers their lifetime milk production potential.

A heifer that has been over-fed will not place well in the show ring either. She will show fat deposits near the tail head and pins, as well as thickness over the withers and in the neck.

Management Timetable For Dairy Calves And Heifers

1. *Disinfecting navel:* As soon as possible after birth, dip the navel cord in iodine or a similar antiseptic.

2. *Colostrum:* Make certain that the calf receives feedings of colostrum from its dam or pooled colostrum from more than one cow within 24 hours following birth.

3. *Removal from dam:* At birth to three days of age, preferably as close to birth as possible, for better management of dam's udder and the calf.

4. *Identification:* Identify by means of ear tag and other means, such as tattoo, color marking sketch, or picture, as soon as possible after birth.

5. *Dehorning:* This is best accomplished when the calf is 10–14 days old.

6. *Removing extra teats:* Rudimentary or extra teats should be removed as soon as they can be distinguished from the four main teats.

7. *Dry feed and water:* Provide calves with a calf starter or equivalent grain mixture within a few days following birth. Start forage feeding and provide water within a week following birth. Provide free–choice mineral feeding as soon as feasible.

8. *Weaning from liquid feed:* Wean at six weeks of age if calves are thrifty and eating at least 1.0 pound of suitable grain mixture per head daily. Continue milk or milk replacer feeding as long as necessary for unthrifty calves.

9. *Removal from individual stalls or pens:* Wait until calf has been weaned from liquid feed for one to two weeks.

10. *Vaccination:* Brucellosis or Bang's vaccination should be done before six months of age. Four months is preferable to avoid positive blood tests showing up at a later age.

11. *Checking growth:* Important periods for checking growth are at 30 days, 6 months, 12 months, 15 months, first freshening, and second freshening.

12. *Breeding heifers:* Plan to have heifers at proper size for breeding at 13–15 months of age. Avoid overconditioned or fat heifers as this may be detrimental to future production.

Recommended Breeding Size for Dairy Heifers

Breed	Body Weight	Heart Girth
	Pounds	*Inches*
Ayrshire	650–700	61–63
Brown Swiss	750–800	64–66
Guernsey	600–650	59–61
Holstein	750–800	64–66
Jersey	550–600	58–60

Source: Extension Mimeo 2556 prepared by B.F. Kelso, Extension Dairy Scientist, Western Washington Research and Extension Center, Puyallup, WA.

Fitting and Training A Dairy Heifer

Having the best animal in the show is only part of the contest when showing dairy cattle. You also have to convince the judge that yours is the best, and this requires a well trained animal that is properly conditioned and fitted.

Preparing a dairy heifer for show begins several months ahead of the exhibit. An early start on both conditioning and training is important. The poorly conditioned animal won't look her best, and one which isn't well trained is impossible to pose correctly.

We should begin looking at the condition of the heifer several months before show date. Assuming the heifer has been fed well and has the needed size and growth for her age, the major concern is making sure the animal is not too fat or too thin. If she is too thin, we increase the concentrate or grain portion of the ration. If the heifer is too fat, we decrease this portion.

Some yearling heifers may require no concentrate to achieve proper condition, while others may need up to five or six pounds of grain concentrate per day. This varies with the animal's size and individual needs, as well as the quality of the forage portion of the ration.

Because hay often makes up most of the ration for dairy heifers, the quality of this forage is quite important. This includes digestibility and energy factors, as well as protein content.

Training the animal to lead should also begin several months before the show, especially for some of the younger showmen. Most dairy calves are relatively gentle and can be trained to lead rather easily if they are penned, fed, and handled on a regular basis.

Training begins with fitting the heifer with a leather or nylon strap halter and tying her to a wooden fence or wall where she can't get away. A solid wall is best to be sure the heifer can't get a foot caught or hurt herself. Feeding some hay on the ground where the heifer is tied will help keep her occupied.

This is different from beef steer training where we usually have a more unruly animal and need a stronger rope halter. Rope halters are not usually needed or recommended for dairy calves.

While she is tied the heifer can be brushed or groomed to get her used to being handled. The animal can be left tied during the day or for relatively short periods if desired. Within two to three days most heifers will be gentle enough to begin teaching to lead.

Training the animal to lead and stand properly should begin several months before the show.

After a few days the young showman can lead the animal to water or feed in a secure area where she can't escape. Later the heifer can be led with the leather show halter. Some calves will resent the chain under the chin on the show halter, and it might be well to remove this portion in the beginning.

After the calf is leading well and can be taken out of the pen into a larger area, the showman should begin teaching the heifer to walk slowly and in small steps. An adult should be there to help and can play the role of the judge while the showman practices leading and posing the animal.

The heifer should be kept in a clean, well-bedded pen when she is not outside and should be penned the last two months before the show. Manure stains are almost impossible to remove, and anything that can be done to prevent them is worth the trouble. Good bedding and a dry pen are the best prevention.

Dairy showmen often turn show heifers out at night to encourage hair growth. In the eastern part of the country some say that rain is also good for improving hair coat; and they go to special effort to turn the cattle out when it is raining. In many parts of the West, we don't know very much about rain.

Show cattle are usually kept inside during the day to prevent sunburning of the hair. Blanketing dairy cattle the last two or three weeks before show has been traditional in many areas. Some showmen prefer not to blanket, believing this discourages hair growth. They want longer hair on the animals to permit a better fitting job through skillful clipping and combing.

Dairy heifers are clipped around the head, neck, tailhead, and tail as well as at particular points on the topline to give the straightness and sharp lines desired. The entire head should be clipped, except for the long hairs on the nose. These should be left because the animal uses these hairs to feel things like feeders and other objects, and they may make the muzzle look wider.

The ears are clipped inside and out, and the neck is also clipped and blended in with the shoulder area. We want the effect of smoothness around the shoulders and sharpness over the withers. Many showmen use the clippers to give this effect and may leave longer hair on top, for combing-up to give a sharper appearance.

Any high points on the topline or around the tailhead are clipped to make a straight topline. The tail is clipped from a point a few inches above the switch up to the tailhead.

For a younger showman, it's nice to have a heifer that is about your size.

Clipping requires practice and experience. An experienced showman should be consulted for advice on clipping individual animals. While this is not a job for a younger 4-H member, the smaller kids can learn by watching and by clipping the less difficult portions. Younger members can also learn by helping clip a practice animal that won't be going to the show.

Much of the clipping can be done two or three weeks before the show, and certain areas, such as the head, should be reclipped a few days before the show. Getting the animals used to the clippers well ahead of the show is a good practice and can prevent that little, last minute panic which sets in near show day.

Washing can begin several weeks before the show and may need to be performed a number of times on hard-to-clean or stained areas. Because washing too often can remove natural oils from the hair, this job should not be overdone. The heifer should be brushed after washing, and brushing on a daily basis when the heifer is dry will help keep her clean and improve the hair coat.

Many kinds of liquid soaps are suitable for washing, as are the special soaps sold for washing livestock. The livestock soaps are made to lather in cold water and may work better for situations where warm water isn't available. All soap must be thoroughly rinsed out of the hair after washing.

It's important to keep water out of the animal's ears. The inside of the ears can be cleaned with a rag dipped in rubbing alcohol to remove wax and dirt.

Showing Dairy Heifers

Fitting, conditioning, and training the dairy heifer is a major part of the preparation for a junior dairy show. But how well you do at the show also depends upon a few last minute preparations and your knowledge of what to expect in the show ring.

Let's begin a few weeks before the show by making sure entries are made and show rules are understood. Even if you have attended this show before, rules can be changed. It's always best to avoid surprises.

When it comes to entries and rules we should never depend upon others to tell us what needs to be done. Your 4-H leader or FFA advisor can be a great help, but in the end it's an individual's responsibility to know the rules. Among the things to check are the health requirements for dairy animals, the need for registration papers, and any suggestions or requirements for clothing worn by showmen.

Dairy showmen wear white in most parts of the country, but this varies with local customs and requirements. Judges sometimes suggest hard shoes be worn for showing rather than soft ones, such as tennis shoes. In these days when many kids don't own hard shoes, this suggestion may be softening up a little; but a mashed foot can still ruin a showman's day as well as his composure.

On the day of the show, dairy animals are fed with the goal of giving them the proper fill. The amount of feed and water given is varied according to the individual animal and the interval between feeding time and show time. Most experienced showmen will feed hay in small amounts and on a continuous basis the day of show to keep the animal eating and to produce the desired fill at show time.

Many showmen also use a bulky feed such as soaked beet pulp or silage to give extra fill on show day. Whether the extra fill is needed depends upon the individual animal and the appearance desired. While cows may eat whatever is put in front of them, heifers will need to be conditioned to any special feeds that will be given at the show.

Water is generally held back for several hours before show to assure the animal will drink before going into the ring. Then the amount of water given is also dependent upon the animal's needs.

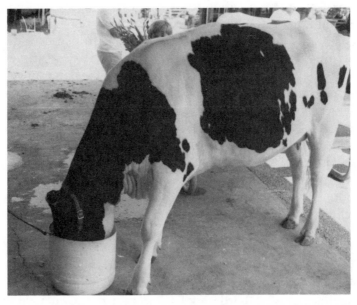

Most showmen like to give show animals a bulky feed, such as beet pulp, to provide extra fill the morning of the show.

All of this may sound vague or complicated to the beginner, but it's really just part of the fun and entertainment that goes with showing livestock.

It is important to know when your class will show and be ready well ahead of time. The last minute clothes changing and rush to the show ring does little for calming the junior showman's nerves — and is somewhat hard on his mother, also. One way to be prepared and keep the white clothes clean is to dress for the show in plenty of time and then wear coveralls over everything until soon before going into the ring. If coveralls aren't available, Dad's oversized work–shirt is a good substitute.

Before entering the ring, make sure the show halter is adjusted to fit the animal properly, and the lead strap is coming out of the left side of the halter (the animal's left). Try to be near the entrance to the ring when your class is called. Know as much as you can about your animal. The judge may ask the birthdate, the sire, or in the case of older heifers, he may want to know if she is bred and the expected calving date.

Many junior showmen are reluctant to be first into the ring, but they shouldn't pass up this opportunity if it presents itself. The first animal into the ring often has the immediate attention of the judge and has an exellent chance to make a good impression.

Dairy animals are shown with the showman walking slowly backward. Practice before the show will help determine the walking speed at which the heifer looks her best. The lead strap should be coiled in the showman's hand but not wrapped around the hand.

The animal's head should be held high. The showman must keep an eye on the judge, but not to the extent of forgetting how the animal is walking or standing. When the judge motions to stop, the showman sets the animal.

Dairy heifers are posed with the rear leg nearest the judge a step backward from the other rear leg. Cows are posed with the leg nearest the judge forward. The beginner may be unable to remember everything or always have the feet in the right position as the judge walks around the ring. For the younger showmen it's more important to have the animal standing in an attractive position, rather than to be constantly moving rear feet.

Most animals can be taught to move the rear feet and stand more correctly when the showman pulls forward or pushes backward on the lead halter. This is where training before the show is important. If the animal is hard to pose at home, it will likely be worse in the show ring.

151

The front feet can be moved by the showman's shoe applying pressure to the animal's foot just above the hoof. This is discouraged by some junior show judges, however, in the belief the kids can move the feet better by forward or backward pressure on the halter. It's best to check local custom and junior show recommendations before telling the kids what to do.

The showman should allow plenty of space between animals when stopping the heifer for the judge or placing her in a line. The judge can see your heifer better if she is not crowded against other animals. Allowing extra space to the front lets you pull up a few steps if the showman behind gets too close.

Showmen should continue showing while the judge gives placings and reasons. Everyone can learn from the judge's reasons regardless of where they are placed in line. Courtesy and sportsmanship are always important in the ring.

Showmen should not be discouraged by their animal's placing. They should learn from the judge's placings and realize this may differ from their opinion or from other judges. That's part of showmanship. Then they should look at their ribbon and be very happy: They've worked hard for it.

A drink of water before showing is usually needed to provide the desired fill and appearance.

FEEDS &
RATIONS

Balancing Livestock Rations

One can always get advice on how to feed his animals. Most of it is good advice, but not all. The junior livestock producer or owner of a few animals is sometimes at the mercy of those who know just slightly more than he does.

The ability to calculate a balanced ration is a valuable skill when deciding which advice to follow and/or in selecting a ration for a particular class of livestock. Calculating a balanced ration is not difficult. All that's needed is a table of nutrient requirements for the animals to be fed and a table of nutrient content for the feeds on hand or to be purchased.

It must be recognized that ration calculation is a mathematical exercise based upon estimates of needs for the animal and usually based upon estimates of the nutrient content of available feeds. Therefore, the calculated ration is an estimate.

Those with more detailed knowledge about characteristics of feeds and with experience in feeding animals are good sources of advice, but it also helps for the beginner to understand the basics of ration formulation. Then, we can better apply what the experts tell us.

When I was a vo-ag student, we were taught to balance rations by the Pearson's Square method. That is still a good way; but unfortunately, most of us can't remember how to use the Pearson Square when we need it.

Pearson's Square works well for calculating the required amounts of two different feeds to obtain a concentrate mix or a ration with a particular protein or TDN (Total Digestible

Nutrients) percentage. It isn't necessary for many simple ration calculations, however.

For simple ration calculations we can use what I call the "guess again" method. With this method one simply finds the nutrient requirements of the particular class of livestock by looking at a table and then locates the nutrient content of proposed feeds in another table.

We estimate how many pounds this animal will eat per day. Then we guess. What if we feed ten pounds of grass hay and four pounds of grain concentrate per day? Will this meet the animal's requirements?

Let's take, for example, a steer weighing 500 pounds. He may eat about 14 pounds of feed per day. This is within the old rule of thumb saying most young ruminant animals will eat about three percent of their bodyweight per day. (Some, such as lambs will eat four percent of their bodyweight or more per day.)

The nutrient table I have handy says this steer needs about 1.5 pounds of total protein and about 7.3 pounds of TDN per day, if we want the steer to gain about 1.5 pounds per day.

When I look at the tag on my feed sack it says this grain concentrate contains 12 percent protein. Looking at the Nutrient Content of Feeds table, I see that "good quality" grass hay contains about 8 percent protein.

If I feed 4 pounds of grain concentrate at 12 percent protein, that's $4 \times .12 = .48$ pounds of protein per day from the concentrate. Ten pounds of grass hay at 8 percent protein will provide $10 \times .08 = 0.8$ pounds of protein per day.

When I add the two together, I see the steer will receive 1.28 pounds of protein per day. My requirements table says this isn't enough protein for this animal. The steer needs 1.5 pounds of protein per day.

To correct the protein deficiency we can feed a higher protein hay, such as good quality alfalfa, or provide a grain concentrate with a higher protein content.

Let's do the same thing with TDN. Total Digestible Nutrients is the commonly used element to estimate total energy derived from a feed. My table says this grass hay contains 47 percent TDN. The 10 pounds of hay $\times .47 = 4.7$ pounds of TDN per day from the hay. Because most concentrates are between 70 and 80 percent TDN, we'll estimate the grain concentrate at 75 percent TDN giving us .75 × 4 pounds = 3.0 pounds of TDN from the concentrate. The two feeds together provide 3.0 + 4.7 = 7.7 pounds of TDN per day. My table says this is adequate. By applying different feeds and estimated nutrient contents we can devise a ration to meet the animal's requirements.

The tables will also help us determine mineral requirements of livestock and estimated amounts of these minerals contained in various feeds. For ruminants the calcium and phosphorus content of feeds and the ratio between these two is often the major mineral consideration, but a mineral mix containing the essential trace minerals is also recommended in most cases.

For swine rations the quality of the protein and the proper amounts of particular amino acids, such as lysine, are important. This is explained more fully in the section of the book on feeding hogs.

Let's remember that nutrient requirement tables and nutrient content of feed tables are based upon averages and estimates. They may contain mistakes, and they are sometimes out of date. Check more than one table before deciding what to believe.

The nutrient content table in a 4–H manual before me says good quality grass hay contains 1.4 percent total protein and 4.9 percent digestible protein. That's obviously a mistake. We have to watch for those.

Nutrient tables may disagree considerably on what good quality grass hay is like. We have to use some judgment. It doesn't help to carry it out to three decimals when we started with an estimate in the first place.

We should also remember that livestock production has changed over the years. Changes in expected performance have increased the recommended nutrient requirements for some classes of livestock. Old tables are often out of date for today's production systems.

One principle to remember when balancing rations is that younger animals will need a higher protein ration than older animals. Lactating animals need more protein than those not milking. So we need to change the ration as the requirements of the animal change with age, pregnancy, and lactation.

A ration exceeding the protein requirement of the animal will not normally cause problems unless this level is carried to great extremes. (I'm not talking about horses. I don't know if this is a problem for them or not.)

Some research with cattle suggests an excess amount of protein supplement changes the balance of micro-organisms in the rumen and reduces the efficiency of digestion for roughages. Too much protein in the ration can cause other problems, such as making breeding stock overfat; but that's an energy problem. The main reason we keep the protein level down is that higher protein feeds generally cost more.

Determining Hay Quality

If our animals could talk, they would surely tell us a few things about the hay we feed them. I think 4-H and FFA animals would have a few words about the flake of hay. They would tell us hay is like snow: no two flakes are alike. The animals would say (if they could), "The book says we need five pounds of hay. It doesn't even mention flakes!"

Because hay forms the ration base for many classes of ruminant livestock, I think the animals would ask us to pay more attention to the amount, type, and quality of hay we feed. The quality of our forage affects all aspects of the feeding program and animal health as well.

I will attempt in this short space to discuss some of the basics of hay quality for those of us with limited experience in livestock production. It will be necessary to generalize, but I'll try to be careful about it.

Most feeding information divides hays into categories, such as grass hays, legumes, or grass and legume mixes. In many areas of the country alfalfa is the most common legume hay, and other types are often compared to alfalfa when discussing quality. Many rations will be based upon "good quality alfalfa."

Hay quality is best estimated by a chemical test in combination with visual appraisal. Small scale livestock producers may be unlikely to perform a chemical test on the small quantities of hay they feed, but a test doesn't really cost much when compared to a few tons of hay. It might be worth the effort in many cases.

Stage of growth at cutting is the single most important factor in hay quality.

Chemical testing of hay is on the increase with commercial haygrowers. Some growers will have chemical test results on all of the hay they sell. Many will be able to tell the buyer the protein content of hay to be purchased.

What does good quality alfalfa look like? Naturally it should be green and leafy. Green generally indicates proper curing and a high carotene content, although some bleaching and loss of color doesn't always indicate poor quality.

The single most important factor in hay quality is the stage of maturity at cutting. Protein content, palatability, and digestibility all decrease as the crop matures. Early cut hay is more desirable in all of these factors. The main indicator of stage of cutting for alfalfa is the amount of bloom present (if any). The number and size of seed heads also indicates maturity in grass hays.

The older books say the best balance of quality with quantity when making alfalfa hay is to cut at 10 percent bloom. Some haygrowers say it's pretty hard to cut 300 acres of hay at just the right time. Growers who sell to dairymen also say, "If my buyers say no bloom, I give them no bloom." For these reasons haygrowers in some areas are cutting much earlier than early bloom.

Two thirds of the protein in alfalfa hay is found in the leaves. Any loss of leaves during curing and baling can result in major losses in quality.

Maybe we should ask, what does good quality alfalfa feel like? Some experts tell us the feel of hay is at least as important as visual appraisal. The softness of hay is related to maturity at cutting, leaf content, and good curing methods. All are important factors in quality.

What type of hay should we feed to various classes of livestock? This is a tricky question. I bring it up to illustrate there are a lot of myths about what is good for certain types of animals.

For example, it has been said that alfalfa is bad for horses. The experts tell us that as a rule this isn't true.

I like the story once told by Dr. William Tyznik, professor of animal nutrition at The Ohio State University. Dr. Tyznik is well known for research on nutrition of horses and gives a lot of talks on that subject. After one of these talks, a lady from the audience approached him and said her horse couldn't eat alfalfa hay because it made the horse's eyes water. She wondered why alfalfa would have this affect.

Dr. Tyznik said, "Ma'am, those are tears of joy. Your horse is just so happy to get that alfalfa that he just can't contain himself."

As a general rule, there is no class of animal that really likes poor hay. We may feed them poor hay if it's cheaper and adequate for that class of animal or sometimes to get more fiber in the diet when using a high grain ration.

It is true that too much good hay can be a problem with some animals because it may cause them to become too fat. Also there is the problem of bloat and other digestive disturbances in particular situations.

How about grass hay? The major differences between grass hays and alfalfa are protein content and much less hazard of bloat from grass hays than from alfalfa. This is sometimes important when feeding high energy rations to cattle.

Although early cut grass hays can have a very good protein content, the protein level declines rapidly as the grass

161

matures. We generally have to add extra protein to rations for young animals when grass hays are a major part of the ration. Mixtures of grass and alfalfa can be very good hays for most classes of ruminants.

Many feeders prefer grass hay over alfalfa when finishing 4–H or FFA steers on a high grain ration. They find there is less danger from bloat and fewer digestive problems with the grass hay.

What is the minimum amount of hay recommended for ruminant rations? For a finishing ration for steers or lambs we generally limit the roughage and increase the grain portion of the ration in order to get more energy into the ration and achieve a faster gain. There may also be times when hay is scarce and prices are high, making it economical to replace a considerable portion of the hay in rations for mature cattle by feeding more grain.

Most recommendations suggest ruminants need a certain amount of roughage in the rumen to maintain proper fermentation. This is variable with age and type of animal as well as what type of grain or concentrate is being fed. When finishing steers, it's not uncommon to limit the amount of hay in the ration to 20 percent of the total.

When finishing lambs on pelleted rations, many feeders feed no long hay. The pellets contain hay, however, and have a fairly high fiber content. In the case of 4–H lambs and steers, we might also notice they are eating straw used for bedding and may be getting more roughage than we realize.

When limiting the roughage portion of the ration for mature cattle or sheep, it would be well to consult a county agent, nutritionist, or other good reference to determine the best course of action with your grains and the type of roughage available in your area.

Management of Small Pastures

A cartoon done several years ago by western cartoonist, Ace Reid, shows a thin cow against a landscape of dirt and rocks, and one cowpoke is saying to the other, "That's why my cattle are heavier than yours. Mine eats rocks!"

I think sometimes we get so excited about breeding bigger and better animals we forget these critters have to eat. In the quest for better breeding we should give at least equal attention to management of the animals. For ruminants, this often includes pasture management.

While the principles of pasture management should be well known to the commercial producer, the part-time farmer or the family with a few 4–H animals may also face some pasture management decisions. One of the first of these decisions is how many animals the pasture can support.

I can safely say nearly all of us who have a few animals have more than the pasture will support. It seems to go with the territory. Our goal should be to make maximum use of the available pasture and feed the animals in drylot the rest of the year.

This may be a good place to explain that the term "drylot" is a bit of a misnomer. Drylot means an area without significant vegetation. In the winter a drylot is often very wet, but it doesn't support much vegetation.

For those of us with a few acres and too many animals, it's generally better to fence a smaller area for a drylot and manage the remainder properly than to overgraze the entire acreage. This is not only better for management of the grasses and legumes in the pasture, but is also better for the animals.

Severe problems with noxious weeds in many areas make an overgrazed pasture a serious liability. When animals are confined to overgazed pastures, they often eat everything except things that might kill them.

Then, after a while, they will go ahead and eat the poisonous plants, too. Animals are better off confined to a small area without significant vegetation, rather than having access to a larger area containing only poisonous weeds.

Proper management of the pasture entails grazing at the right time for the grasses or legumes present and having animals off long enough for the plants to make sufficient regrowth — both between grazings and between seasons. Management may also include fertilization or reseeding.

Many commercial outlets sell pasture mixes which include a large number of grass species, on the theory that if you plant enough different species, something is bound to grow. If the buyer doesn't know something about the species desired, it's not uncommon to get a mix containing a lot of things you don't really want.

Aside from paying for grass species you don't want, a second problem with seeding several different species is that each grass has different growth habits and management requirements. If these aren't compatible, it's difficult to devise a grazing schedule to fit the species present.

There is a third problem. The livestock will graze the most palatable grasses first and hardest; and if the pasture is grazed heavily, the grass that survives will be the one the animals like least. It's best to get recommendations before buying the pasture mix.

Then there is the question of fertilization. Again, it's best to get recommendations for the area and species present in the pasture. For nearly all small acreages, fertilization of grass pastures will pay off.

In addition to the increased grass production achieved from fertilization, the application of nitrogen fertilizers also helps greatly in controlling undesirable broadleaf weeds. The nitrogen helps the grasses more than the broadleafs, giving the grass a boost in competition with the weeds.

Fertilization is necessary to achieve top production of grass
pastures. In some areas of the country irrigation is also critical.

If legumes are a major part of the pasture, application of
nitrogen fertilizers can be detrimental to the legume stand.
These fertilizers will cause an increase in grass growth and
thereby decrease the stand of legumes over a period of years.
The benefits to the grass production are often worth the
reduction in the legume stand. Amounts and kinds of fer-
tilizer for various types of pastures should be determined
from local recommendations.

Whether pastures should have legumes and the desirable
ratio of legumes to grasses depends upon the type of use the
pasture will have and the animal species to be grazed. For example, in many areas it is recommended that breeding
pastures for sheep not contain legumes , especially clovers.

In other areas it is common to graze legume pastures
(including clovers) before and during the breeding season. It's
best to get information and advice from those familiar with
your area and growing conditions.

The best sources of information on grasses and legumes for
pasture seedings are your local Soil Conservation Service office and county extension office. It's important to seed grass

and legume species that are adapted to the area as well as the use or management system you have in mind.

A common recommendation is to strive for a stand of less than 50 percent legumes in a pasture to be grazed by cattle or by sheep during periods outside the breeding season. A lower level of legumes is normally expected to lessen the risk of bloat.

Many researchers will admit bloat is a complicated and poorly understood process. In one older agronomy textbook the author writes that in conversations with several animal researchers, he found that techniques used by one researcher to reduce the incidence of bloat were being used by another to produce this condition for study.

Feeding Wheat

Although wheat is not usually considered a feed grain in some parts of the country, there are many situations where this grain becomes a viable feed alternative. Wheat may be an economical feed because of a poor wheat market in general or because of reduced quality from contamination with other grains or sprout damage.

Sometimes transportation and selling costs in the wheat growing regions make it more economical to feed wheat on the farm than to haul it to town and transport other grains back to the farm. This can be overdone, however.

I'll always remember the rancher at a nutrition conference some years back who was feeding wheat when the price was over $5 per bushel and corn was about $3. When an incredulous Extension Specialist asked why he was feeding that expensive grain, the rancher replied, "I already have the wheat."

While this man had become used to feeding wheat, many folks are not, and there are a lot of scarey stories about wheat being a "hot feed." If I could draw cartoons, the first one would show a rancher standing beside a charred steer carcass with a puff of smoke curling skyward; the rancher would say to his wife, "It looks like that new ration may be too hot for these steers."

It pains me to recall how many times I've heard 4-H and FFA members told a particular feed would burn-up their animals. I figure you can make them sick, and you can kill them; but you can't burn them up.

Wheat can provide a significant portion of the grain ration for all types of livestock and can be the sole grain fed to some classes of animals. This grain may be fed as the only energy source for poultry and swine if properly supplemented. Because it can cause rumen disorders if fed at high levels, it is generally recommended that wheat make up less than 50 percent of the ration when fed to cattle and sheep.

One consideration when advising 4-H or FFA members on feeding any grain (or hay) is to make sure they weigh the feed rather than measure it. A coffee can full of whole wheat weighs nearly twice as much as the same can full of whole oats. If the feeds are ground, the weight-per-can changes; and when you buy a new batch of grain, the weight may change again.

When my old friend Joe Johnson was a Washington State University Extension Specialist he reportedly said while interviewing members for state 4-H awards, "The first kid that comes in here and can tell me how much a flake of hay weighs is going to Chicago as the State Horse Project Winner!"

It's too bad, but I don't think Joe ever found the kid. My experience would say a flake of hay weighs between 2 and 45 pounds, depending upon type of hay, conditions of baling, and what time baseball practice begins.

When wheat is fed as less than 50 percent of the grain ration, most studies show it to be worth about the same as shelled corn for feeding steers. Barley is generally considered to be worth 90 percent of the value of wheat for feeding cattle. Coarse grinding is recommended for wheat fed to cattle.

Wheat is a good energy source for market lambs or sheep breeding stock and grinding is not required for either lambs or breeding sheep (except for old ewes that may be short of teeth.) Although some studies have shown wheat can be fed as the total grain ration for market lambs, most experts recommend mixing with barley or oats so that wheat makes up less than 50 percent of the ration.

Because of the higher energy content of wheat and the potential for mistakes in measuring mentioned earlier, this grain may not be quite as foolproof for feeding by kids as something like barley, oats, or prepared pelleted feeds.

However, a little caution can prevent "burning them up," as they say.

It's important to feed enough roughage when feeding wheat to lambs or cattle. It's not like the pelleted feeds which contain a percentage of hay.

Wheat is an excellent feed for swine and has performed even better than corn in many studies. Some trials have shown better feed efficiency from pigs fed wheat than those fed corn, although the rate of gain may be slightly less for those on the wheat ration.

The protein content of wheat is higher than corn; but when fed to swine, supplementation is required for added protein and for amino acids such as lysine. There is a great amount of variability in the protein content of different types of wheat, with the hard red varieties normally providing more protein than the common soft white varieties.

It is generally recommended that wheat be coarsely ground or pelleted for swine rations, but it can also be fed whole. If whole wheat is available at a much lower cost than coarsely ground or rolled wheat, a slight increase in feed efficiency from grinding may not offset the increased cost of this operation.

If a youngster has only a few pigs and access to the wheat in Dad's bin, this whole wheat may be preferable to the more expensive rolled wheat available in town. Some references suggest that whole wheat be self-fed to permit time for the pigs to chew it, as they tend to gulp it down when hand fed. Many pig feeders soak wheat in water overnight on the theory that this improves digestion of whole kernels.

What about the sprouted wheat that sometimes results when poor weather at harvest causes the grain to sprout in the field? A University of Idaho study showed wheat containing up to 60 percent sprouted kernels to be nearly equal to sound wheat for swine rations. Feed efficiency was slightly less for pigs fed sprouted wheat, but their rate of gain was equal to those fed sound wheat.

Research at Washington State University on feeding sprouted wheat to beef cattle showed no difference in performance or carcass characteristics for cattle fed wheat with up to 58 percent sprouted kernels, as compared to cattle fed sound wheat.

WHEAT RATIONS FOR FINISHING CATTLE

In the following rations wheat makes up 40% of the total ration, when the roughage portion of the ration is also considered. These rations are formulated for roughage (hay) to be fed as slightly less than 15% of the total ration (example: 16 lb. grain with 2.4 lb. of hay).

Ration 1.   %
    Wheat............................46.5
    Barley.............................46.5
    Molasses.......................... 5.0
    Ground Limestone................. 1.5
    Salt (trace mineralized).............. 0.5
To be fed with alfalfa hay

Ration 2.   %
    Wheat............................46.5
    Barley.............................43.0
    Molasses.......................... 5.0
    Soybean Meal..................... 3.5
    Ground Limestone................. 2.0
    Salt (trace mineralized).............. 0.5
To be fed with grass hay

Add vitamin A to both the rations at the level of 1000 IU per pound of feed.

Source: Washington State University Extension Bulletin EB1317 *Feeding Washington Wheat*

WHEAT RATIONS FOR FINISHING LAMBS

Ration 1.   %
    Wheat............................30.0
    Barley.............................29.5
    Salt (trace mineralized).............. 0.5
    Alfalfa hay........................40.0

Ration 2.   %
    Wheat............................30.0
    Barley...........................39.0
    Ground limestone................... 0.5
    Salt(trace mineralized).............. 0.5
    Alfalfa hay.......................30.0

Ration 3.   %
    Wheat............................30.0
    Barley...........................32.0
    Soybean meal...................... 6.0
    Ground limestone................... 1.5
    Salt (trace mineralized)............. 0.5
    Grass hay........................30.0

Source: Washington State University Extension Bulletin EB1317 *Feeding Washington Wheat.*

WHEAT RATIONS FOR PIGS

Ration 1.Starter, 20–50 lbs., 18% protein

                                                  %
    Wheat............................72.2
    Soybean Oilmeal(47.5% protein).......25.2
    Limestone......................... 0.8
    Dicalcium phosphate................ 0.9
    Vitamin premix.................... 0.5
    Trace mineral premix............... 0.1
    Salt.............................. 0.3

In the following rations cull peas, a protein feed available in the Northwest, are substituted for a portion of the soybean oilmeal to lower the cost of the rations.

Ration 2.Grower-finisher, 80–130 lbs.,14% protein

                                                  %
    Wheat............................77.6
    Soybean Oilmeal....................10.0
    Cull peas.........................10.0
    Limestone......................... 1.0
    Dicalcium phosphate................ 0.5
    Vitamin premix.................... 0.5
    Trace mineral premix............... 0.1
    Salt.............................. 0.3

Ration 3. Grower-finisher, 130–220 lbs., 13% protein

	%
Wheat	79.5
Soybean Oilmeal	6.0
Cull peas	12.0
Limestone	1.1
Dicalcium phosphate	0.5
Vitamin premix	0.5
Trace mineral premix	0.1
Salt	0.3

Ration 4. Grower-finisher, 130–220 lbs., wheat plus synthetic lysine

	%
Wheat	96.80
Lysine	0.35
Methionine	0.10
Limestone	1.20
Dicalcium phosphate	0.70
Vitamin premix	0.50
Trace mineral premix	0.10
Salt	0.25

Source: Washington State University Extension Bulletin EB1317 *Feeding Washington Wheat*

JUNIOR
SHOWS

Winning and Losing

Those who hang around fairgrounds each summer are familiar with the cry of the discouraged parent: "This is the last year we are going to do this!"

Maybe I should say the lie of the discouraged parent. Of course, this is not the last year we are going to do this; we're just threatening the kids.

Along with the fun and excitement of livestock shows, there are also pressures for both parents and kids. One of the common frustrations for the kids is deciding how well they did. If the goal is to win, some adjustments are needed when one learns he didn't.

Each of us has the almost daily opportunity to win or lose at something; however, only on special occasions do we have the chance to do it in front of a crowd.

For many youngsters and some adults the special occasions for winning and losing are sporting events and livestock shows. Each year thousands of 4-H and FFA members will be winning and losing in front of the crowd at county fairs across the country.

I think the psychologists would tell us that winning and losing is important to all. Obviously, people who lose all of the time don't adjust well to life's demands; but they say those who win all of the time don't adjust well, either. Some experience at both winning and losing is important for each individual.

I believe winning and losing is an individual thing. Through some combination of experience, social environment, and

175

maybe heredity, each person decides what constitutes victory or defeat.

Most readers have seen the youngster, with the animal at the bottom of the class, break into a big smile when presented a white ribbon. We may have expected this young person to be disappointed at receiving what many consider to be a third rate prize. But she felt good about her efforts and the ribbon was important.

Then, we have seen the boy or girl in second place frown when he or she received a fancy rosette and a blue ribbon. Sometimes it's hard to tell the losers from the winners, isn't it?

Some are fortunate and can feel good about a blue, red, or white ribbon, if they worked hard and did their best while others are satisfied only by first place. The latter group is not likely to be satisfied for very long.

The kids develop certain expectations; and as they grow older, we see all sorts of reactions to winning and losing. We see all sorts of reactions from their parents, too.

This young lady has fun showing cattle. That's the way it should be.

Some of the reactions seen at livestock shows lead many adults to argue that competition is bad for kids while others staunchly maintain that all of life is a competition. The most fervent competitors believe when the world comes to an end there will be one who survives; and that one shall be known as the winner.

I don't know who's right, but I'm a strong supporter of competition, as long as I can win. I think most people feel that way. We just have a little trouble deciding who's winning.

While hard work and dedication is healthy, being first at all costs is not and often leads to cheating in its many forms.

I'm not saying youngsters should be satisfied with less than their best effort. However, it's important for parents to keep things in perspective. We should emphasize the effort rather than the result.

We have all seen entire classes of young showmen disappointed when the judge gave only a few blue ribbons and other groups pleased when the judge gave blue ribbons to the entire class. It may be hard for the kids in the first instance to understand that a red ribbon may indicate good effort or those in the second case to realize that the blue ribbon may not indicate much of anything. This is something I wouldn't tell a youngster, but which can be very true.

If you've read this far, you may be willing to read an old story I've told a few times. It's true, too. Years ago at a county fair a little girl ran up to tell me, "I just showed my lamb, and I got a ribbon. And I show it again this afternoon for showmanship."

I started to ask, "What color ribbon did you get?" but managed to hold my tongue. It occurred to me that the girl was happy with her effort and with her lamb, and the color of the ribbon was really not that important. Besides, it was none of my business.

I later learned that this little girl's lamb was awarded a white ribbon and stood near the bottom of the class. But that was the judge's opinion, which need not have a big impact on either the girl or the lamb.

Did she learn anything? Sure. She came to the fair the next year with a much better lamb and eventually made enough

177

money on her lambs to buy a horse — which is a lot more fun than a lamb.

What's the moral to the story? I think as adults we put too much emphasis on what color of ribbon the kids receive and tend to emphasize the result rather than the effort.

As a result the youngsters are often happy with a poor job that nets a blue ribbon and disappointed with a good job that produces a red. The judge may compensate by awarding 80 percent blues to avoid too much disappointment.

And what about trophies? Everybody likes trophies! No one should be disappointed if he doesn't get one, though.

The adults don't have to take all of the blame. The kids form many of their views without help from us.

I guess I'm suggesting that as we go through this show season, we consider our actions and our words in relation to what the young showmen are supposed to be learning from their experiences.

It won't be easy; but considering the pressures and the circumstances, I think we parents do a remarkably good job of keeping our heads on straight.

Skinning Cats

When I was a kid (several years ago), people would often say, "Well, that just shows there's more than one way to skin a cat." This useful saying may not make much sense to the youth of today. It really wasn't very profound when I was a kid, either, but we still hear it occasionally.

Few people have skinned a lot of cats, and most of us would be hard pressed to demonstrate a single method, let alone more than one. However, the saying is still good, because it fits so many situations, and the meaning is obvious (in a roundabout way).

I was recently reminded there is more than one way to skin a cat as I watched a filmstrip on fitting and showing pigs. One of our 4-H club's leaders obtained the film from the vocational agriculture department, and we were showing it to the club.

It was a good film and very well done, but it contained a few suggestions contrary to what is commonly recommended in many areas of the country.

For example, while many areas are discouraging the use of artificial powders and oils on show pigs, the filmstrip demonstrated how to coat a white pig with a mixture of powder and water to create a mud pack effect. Later brushing took out nearly all of the powder, leaving the slickest, whitest pig you can imagine. (Sort of a meat packer's nightmare, from what they tell me.)

Midway through the filmstrip, one of the boys with some pig showing experience crawled across the living room floor to where I was operating the projector and whispered, "They're showing some things we can't do."

"Yes, they are," I replied, "and after the film is finished we'll have you talk about some of those things." And we did.

Don't get me wrong; this was a good visual aid that provided a chance to point out some important ideas for fitting pigs for show.

At first I concluded the film was old, and these were outdated methods. Then I realized, maybe the film is new, and I'm old! Finally, I thought, "Well, that just shows there's more than one way to skin a cat."

Doing your best is what it's all about.

The lesson here for me is that we sometimes become awfully excited about teaching the kids the correct way to fit and show animals. We have them clip the hair to a certain point on the pig's tail (if the pig has a tail). We have them comb the steer's hair up, then we have them comb it back down; and then we put a topknot on its head. And we teach the lamb showmen how to do figure 8's, just in case the judge is big on that maneuver.

Some would say we should spend more time talking about how to get twin lambs and keep them from dying, rather than doing figure 8's with the ones that are still alive.

What are we trying to accomplish with all of these fitting and showing methods and maneuvers? Well, I would expect we all agree on what we want to accomplish, but it's important to stop and remember that what everyone knows is correct today was wrong yesterday and will probably change tomorrow.

So where was I? Oh yes, watching a filmstrip showing the correct way to fit and show a pig at a particular place and time and thinking we need to keep this fitting and showing practical so that we don't have to debrief the kids after each show. That way they can learn some basic principles that apply everywhere.

We also need to keep things simple because each time we get a bit fancy we find fewer and fewer kids doing their own work.

I'm thinking maybe we don't have to clip a junior steer quite as nicely as the open showman would. Or maybe a 4–Her can clip hair from the pig's ear with a pair of scissors rather than electric clippers. (A little stubble is not a big thing to a pig.)

I'm reminded of a little league coach several years ago who was giving the kids a pep talk before a game. He said, "We're going out there to win this game; winning the game is what it's all about!"

"No, it isn't," objected my wife, sitting beside me on the bleachers.

"Don't argue with the coach, dear," I cautioned.

181

Thoughts About Showmanship

Nearly everyone will agree showmanship classes are important for junior livestock shows. These classes recognize the youngster who works hard, puts forth the best effort, and does the best job of preparing his or her animal for the show. Right?

Some would say, "Well, not always. It depends upon what the judge wants." So the question again raises its ugly head, "What does the judge want?"

Does the judge want a tuft of hair on the steer's head? How much hair should we clip on the pig's tail, and do I have to wear a white shirt? Will I win if I spit out my chewing gum?

It's hard to tell exactly what the judge may want. I have sometimes thought if we could do that we wouldn't need a judge. Instead, we could just make up a score sheet and check things off as they are accomplished.

This is an exaggeration of course, but we have to give the judge some room for judgment.

I think in most cases the judge wants to provide a good test of the showman's ability, wants the kids to learn from their efforts, and wants to be practical.

This is not always easy. Sometimes it's hard to remain practical and also provide a test of showmanship ability. If the kids are all set to do figure 8's with their lambs, someone will be upset if they don't do some.

Last year my kids' club had a meeting a few weeks before the county fair to learn about fitting and showing their lambs. We had hoped to teach them how to do figure 8's but found none of us leaders or parents could remember how to do the crazy things.

It was a little embarrassing for awhile, but most of the lambs were in the early stages of training, and somersaults seemed more popular anyway. So we promised the kids we'd get back to them later.

I have sometimes thought showmanship is in danger of becoming too artificial, and I know others who feel the same way. However, few people will voice this criticism publicly, probably because most are smarter than I.

Maybe showmanship doesn't have to be totally practical. There's nothing practical about tossing a ball in a basket with no bottom, but we do that and call it basketball. People get awfully excited about basketball, too; and we think kids learn something from playing it.

Showmanship probably teaches more than we give it credit for. There is definitely some sportsmanship involved and cooperation between members in preparing and showing the animals.

Good sportsmanship is always important.

Learning to handle the project (and maybe do figure 8's) requires learning about the animal's habits and how it reacts to the handler. If fitting and showing weren't necessary, some kids would not learn nearly as much about care and grooming.

Then there is the competition aspect. I think most kids like to compete because this provides a chance to succeed. They may not care about beating someone else, but being in the game can be important in itself.

Showmanship is an opportunity for learning more about project animals than would be learned otherwise. It's a chance to feel a sense of accomplishment in having a well-groomed animal; and it's a chance to win, or succeed, or just get in the game.

It is true, of course, that some showmen don't learn as much about fitting and grooming as they should because a parent or Grandpa does it for them. This is a problem of emphasis on winning rather than learning from the project, or sometimes it's just an inability of an adult to stand by and watch what the kids are doing to those animals.

Younger members need help and instruction, but we shouldn't expect perfection.

There are a lot of gimmicks in preparing animals for show; but when everything is considered, I believe most of these tricks are given far more credit than they deserve.

I heard a junior fair judge many years ago comment to an audience, "The boy in third place has a beautifully fitted lamb. And I'm not accusing anyone; but if he did that work himself, he certainly must be a professional."

This man was judging a class of beginning lamb showmen, and his comment was a direct expression of what many have thought but declined to say into a microphone.

This judge was not invited back again for several years, but we had him back years later after things cooled down a bit.

He was absolutely right. There isn't a first year 4-H member in the country who can fit lambs the way that little boy's mother can.

Junior Livestock Sales

Few events have been the subject of more applause, criticism, and general debate than the junior livestock sale. Sometimes it's difficult to keep the sale and its purposes in perspective.

Although I have no particular position or authority to explain junior livestock sales to others, I do have some opinions. All of my opinions come with a disclaimer — they may be changed or altered at any time as a result of thinking, aging, or being beaten into submission.

As far as I can tell, the purpose of the first junior livestock sale was to get rid of the animals. I have been told this was necessary to prevent diseases from being taken back to the farm with market livestock from the show. That doesn't make total sense to me, but it sounds OK for awhile.

I believe the second purpose of the sale is to give the kids some money and to reward their work. Buyers give the kids some money to help them pay for project expenses, save money for college, or maybe buy a new motor-scooter. Different buyers have different reasons.

I was talking to a friend recently who said he has four sheep. "What are you doing with four sheep?" I asked.

"I bought them from the grandkids," he said. "Whenever somebody needs some money, I buy a cow or a sheep or something."

People like to give the kids some money occasionally. That's why we have scholarships.

A third purpose of the sale is to support the junior livestock program (4–H and FFA). This may be somewhat confusing because little if any of the money spent in the sale goes to the total 4–H or FFA program in most cases.

However, people who buy the animals want to support the kids and their projects; this is their way of supporting the junior livestock program. I'm sure there are other reasons to have a junior sale, but that's a start.

Now, how can the junior sale be improved? I believe we need to keep our purposes in mind if we want to make improvements.

If furnishing animals of the desired weight to the meat packer is not a major purpose of the sale, I don't get excited about changing sale rules to get two more cents per pound from the packer.

I'm not just saying that to be nasty but to point out that while there are reasons for all of the things we do in an attempt to improve a show or sale, it's very easy to lose sight of our goals when we start making rules.

If selling rabbit projects through the junior livestock sale meets all of the sale's purposes, why not sell the rabbits? Many sales are now doing just that.

It would be nice to make the junior sale as educational as possible, but it's not easy. I'm not always sure what we should teach the kids.

Some of the most successful livestock producers today never learned to sell animals the way everyone else does. These people came up with a better way.

If a major purpose of the sale is to give the kids some money, it's important to encourage buyers and make them feel good about the investment they are making.

Thank-you letters from the kids are very important, and many clubs or sale committees furnish pictures of the kids and their animals with certificates for buyer recognition. These activities not only encourage buyers for the sale but also generate good feelings and support for the youth program.

One county prints a booklet showing that buying animals through a junior livestock show can be a good way to purchase quality meat.

Some clubs take pictures of project animals before the show.
The picture can then be presented to a junior livestock sale
buyer with a thank you note.

When I was a young county agent many years ago, I had
thoughts like other young county agents. I sometimes thought
the livestock sale drew so much attention and time that it
detracted from the total 4-H program. Many people also feel
this way about the junior livestock program or possibly the
horse program.

Then I began to notice over the years that no matter how
much we ignore something that is successful, we don't seem
to make any faster progress with the things that aren't as successful. I began to realize that things like junior livestock
sales, horse programs, and style shows help generate en-
thusiasm and support for lots of other things.

This is called the "rising tide theory" in economic circles. It
says we're all going up together, or we're all going down
together.

So now as a parent and somewhat disorganized 4-H leader,
I tend to say, "If it ain't broke, don't fix it." The junior
livestock sale is a good vehicle for increasing support for
junior livestock programs as well as other youth programs.

187

Using Carcass Data

Collection of carcass data has become standard practice for many junior livestock shows during the past 20 years. Besides being an excellent educational activity, the collection of carcass data over a period of years provides a chance to look back and see changes that have occurred in the livestock industry.

Like many things that have been around awhile, we tend to take carcass data for granted and probably don't give it enough credit for improvements in livestock breeding. As with many things we do in junior livestock shows, the parents have learned a lot more from livestock carcass data than the kids have.

But, the kids learn from this information, too; and it's usually up to the parents or club leaders to explain what a ribeye muscle or backfat measurement has to do with carcass evaluation. It's easy for those in the livestock business to forget that many 4–H or FFA members and their parents have never heard these terms before.

I'm not in the livestock business, but I have looked at more than a few of these data sheets and will attempt to describe some of the terms included in carcass programs and a few changes that have occurred over the years. I won't attempt to describe requirements for carcass awards, as these vary among programs.

In the case of steers, one element of change is carcass weight. Not too many years ago a steer with a 700 pound carcass was considered a big animal. Today carcass weights for many shows average well over 700 pounds.

Carcass weight for show steers is normally about 60 to 62 percent of the animal's liveweight, but this figure varies considerably. This percentage (carcass weight divided by liveweight times 100) is called dressing percentage.

One of the early goals of carcass evaluation programs for steers was to produce animals with less fat. If we compare data from 20 years ago with today's, we'll see that today's animals have considerably less fat per 100 pounds of carcass.

Many carcass evaluation programs require steer carcasses to have less than .08 inch of fat cover per 100 pounds of carcass weight to qualify for recognition. Most modern-type cattle from junior shows will have considerably less fat cover than this.

I know someone will say, "Yes, but a lot of them don't grade choice." It is hard to have everything, isn't it?

Another early goal of carcass programs for steers was to produce animals with a higher percentage of muscle in the carcass. Total muscle is indicated by measurement of the ribeye muscle at a cut between the twelfth and thirteenth ribs.

If we look back at the old data, we'll find this hasn't happened. As steers got bigger, the ribeye measurement has become a smaller percentage of the carcass weight. This has resulted from a change in the style of cattle we now produce.

It's just more difficult to produce an 800 pound carcass with a 16 inch ribeye than it is to produce a 500 pound carcass with a 10 inch ribeye. I think it's more of a mathematics problem than a breeding problem.

A full description of meat quality grading would require more explanation than appropriate for this section, so let's just say that beef quality grades are determined by the amount and distribution of marbling (small fat deposits) in the ribeye muscle, meat color and texture, and indicators of the animal's age and sex. For junior show steers, the meat color in the instance of dark cutting carcasses and lack of marbling are the most common deficiencies for getting into the choice grade.

Lamb carcass weights have also increased in recent years and will average around 60 pounds in many shows. The dressing percentage for lambs is normally about 50 percent, but

189

varies considerably. With live weights of 115 to 120 pounds, we would expect about a 60 pound carcass.

Dressing percentages have long been a point of discussion among livestock producers and buyers but aren't normally considered in carcass data programs. There are many factors involved in the dressing percentage of individual animals; but when everything else is equal, a fatter animal will normally have a better dressing percentage.

For this reason, it's possible to have animals with a high dressing percentage and then have to throw much of it away as trimmed fat in the case of steers or to market it wrapped around a lamb chop in the case of lambs.

Fat cover on lamb carcasses is measured over the ribeye muscle and to the side of the backbone. A gauge is generally used for this measurement because lamb carcasses aren't cut for grading, as steer carcasses are.

Internal fat or kidney and pelvic fat is a second category describing trimness for beef and lamb carcasses. This may be weighed in some evaluation programs, but it is generally a visual estimate by the grader instead. This category is omitted in some junior show carcass programs.

Total muscling in lambs is indicated by leg score when loin eye area can't be measured. Leg score is a visual appraisal by the grader and is indicated by numbers such as, 15 — high prime, 14 — prime, 13 — low prime, 12 — high choice, etc.

Some really good lamb and swine carcass programs obtain more information by cutting the carcasses and taking more weights and measurements, but this often isn't possible in a commercial plant.

Carcass evaluation for swine has changed with revisions in U.S.D.A. pork carcass and slaughter hog grading standards. The preliminary carcass grade is based upon a combination of backfat thickness measured over the last rib and a visual appraisal of muscling (muscling score).

A muscling score of "1" indicates thin muscling; "2" is medium muscling; and "3" is heavy muscling. The information I have states a thin muscled hog cannot be graded No. 1, and thick muscling improves the grade by one full grade.

Details of swine grading changes are available from the National Pork Producers Council or at county extension offices.

APPENDIX

Glossary

amino acids- the basic constituents of protein.

antibody- various substances in the blood or developed in immunization which counteract toxins or bacterial poisons in the animal's system.

barrow- castrated male swine.

blackleg- a disease of cattle, often fatal, characterized by gangrenous swelling of the upper parts of the leg.

bloat- condition in ruminant animals caused by excess production of gases in the rumen.

boar- uncastrated male swine.

Bo-Se- pronounced Boe-C; injectable selenium used to prevent or correct selenium deficiency.

brucellosis- any of a variety of infectious diseases caused by a parasitic bacteria, causing abortions in animals; also called Bang's disease.

breeder- often denotes someone who raises purebred stock; in the case of individual animals the breeder is the owner of the animal's dam.

bummer lamb- orphan lamb or one that must be fed artificially.

BVD- abbreviation for *bovine virus diarrhea*; disease of cattle characterized by diarrhea, loss of appetite, and fever.

castrate- removal of the testicles from a male animal.

colostrum- the first milk secreted by the animal, containing a large amount of protein and immunizing factors for the newborn.

coccidiosis- a disease in which the intestines are infested with a protozoan organism called coccidia.

conformation- the general shape of the animal as determined by his framework or skeleton and muscle structure.

concentrate- the high energy portion of a ration; usually made up of grains, or grains and a high protein source, such as soybean oilmeal.

complete feed- feed that can be fed alone; some pelleted feeds would be examples.

creep ration- ration provided to very young animals; usually fed in a location excluding access by larger animals by means of panels or other structures with small openings.

crossbred- an animal whose sire and dam are of different breeds.

dam- female parent.

dressing percentage- the carcass weight divided by the live weight × 100.

drylot- a relatively small area supporting little or no vegetation where animals can be confined.

enterotoxemia- pulpy kidney or "overeating disease" caused by the sudden release of toxin by the bacteria *Clostridium perfringens* Type D in the digestive tract of sheep; highly fatal; vaccine available.

enzootic abortion- in sheep characterized by premature birth and abortion.

ewe lamb- ewe less than one year old.

feeder pig- pig that has been weaned and is ready for feeding; usually refers to pigs of 30 to 50 pounds; in junior livestock shows, may refer to any pig less than market weight.

fiber- (in the diet) most fiber is made up of cellulose which can be broken down by the bacteria in the stomachs of ruminant animals (e.g., sheep, cows) into useable carbohydrates.

fill- the amount of feed and water in the animal.

flake of hay- what the kids tell you they fed the animals; something less than a bale.

flushing- increasing the nutrient level or quality of the ration prior to breeding in an attempt to increase ovulation; a common practice in sheep and swine production.

free-choice- providing feed, water, or minerals in such a way the animals can eat or drink whenever they wish.

gestation- the period of pregnancy.

gilt- young female swine.

grade- animal which has one purebred parent but is not eligible for registration.

grain- seed of the cereal grasses, such as corn, barley, oats; often refers to the concentrate or high energy portion of a ration.

hand feeding- feeding animals a specific amount on a regular schedule, such as twice a day.

heat cycle- period of time in which a female animal may be bred and conceive.

heart girth- circumference of the animal measured just behind the front legs.

heifer- young female bovine.

ketosis- pregnancy disease resulting generally from inadequate nutrition for the ewe the last three to four weeks of pregnancy.

lactation- period during which the female is producing milk.

legume- plants of the family with fruit that splits on two seams, such as alfalfa, clover, peas.

marbling- fat deposition within the ribeye muscle; a factor in determining carcass grade for steers.

packer- person in the business of slaughtering live animals and selling meat wholesale.

parturition- giving birth.

Pearson's Square- method of calculating ratios of two ration ingredients in order to meet desired nutrient requirements.

pregnancy toxemia- see "ketosis."

protein-(crude)-all the nitrogen–containing compounds in a feed; (digestible)-approximate amount of true protein in feed.

protein supplement- feed of high protein content which is added to increase the protein level of a ration.

prolificacy- production of numerous offspring.

purebred- animal which is registered or eligible to be registered.

ration- everything available for ingestion by the animal; includes concentrates, hay, pasture, browse, etc.

registered- animal which is recorded with a recognized breed association.

ribeye muscle- muscle located along both sides of the backbone; used in determining quality grade and calculating cutability in beef slaughter animals.

rolled grain- grain which has to be processed with a roller mill.

rumen- the largest compartment of the ruminant animal's stomach in which the major roughage utilization occurs.

self-feeding- making feed available to the animals at all times.

shot- an injection.

showmanship- ability to show; a class in junior livestock shows which judges the ability of the exhibitor to prepare and show the animal in the ring.

sire- male parent.

steer- castrated male bovine.

tags- matted wool; separated from fleece and lamb's wool when shearing.

TDN- Total Digestible Nutrients; commonly used as a measure of energy in a livestock ration.

vaccination- introduction of antibodies into animals causing them to produce an immunity or tolerance to a disease.

vibriosis- in sheep and cattle characterized by abortion in late pregnancy or birth of dead or small fetuses.

weaner pig- term used in some areas to denote a recently weaned pig, usually 25 to 40 pounds.

weaning- separation of the young from the dam.

wether- castrated male sheep.

withers- area over the top of the shoulders on dairy cattle.

Registry Associations

BEEF

Angus	American Angus Association 3201 Frederick Blvd St. Joseph, MO 64501
Beefalo	American Beefalo World Registry 116 Executive Park Louisville, KY 40207
Beefmaster	Beefmaster Breeders Universal 6800 Park Ten Blvd, Suite 290 West San Antonio, TX 78213
	Foundation Beefmaster Assn 200 Livestock Exchange Bldg Denver, CO 80216
Braford	International Braford Assn PO Box 2727 Ft Pierce, FL 33454
Brahman	American Brahman Breeders Assn 1313 La Concha Lane Houston, TX 77054
Brangus	International Brangus Breeders Assn 5750 Epsilon San Antonio, TX 78249
Charolais	American International Charolais Assn PO Box 20247 Kansas City, MO 64195
Devon	Devon Cattle Assn PO Drawer 628 Uvalde, TX 78801
Gelbvieh	American Gelbvieh Assn 5001 National Western Dr Denver, CO 80216

Hereford	American Hereford Assn PO Box 4059 Kansas City, MO 64101
Limousin	North American Limousin Foundation 100 Livestock Exchange Bldg Denver, CO 80216
Maine-Anjou	American Maine-Anjou Assn 567 Livestock Exchange Bldg Kansas City, MO 64102
Pinzgauer	American Pinzgauer Assn Rt 1, Box 104 E Kelley, IA 50134
Polled Hereford	American Polled Hereford Assn 4700 E 63rd St Kansas City, MO 64130
Red Angus	Red Angus Assn of America 4201 I-35 North Denton, TX 76201
Red Brangus	American Red Brangus Assn PO Box 1326 Austin, TX 78767
Red Poll	American Red Poll Assn PO Box 35519 Louisville, KY 40232
Salers	American Salers Assn 101 Livestock Exchange Bldg Denver, CO 80216
Santa Gertrudis	Santa Gertrudis Breeders Int'l PO Box 1257 Kingsville, TX 78363
Shorthorn	American Shorthorn Assn 8288 Hascall St Omaha, NE 68124
Simmental	American Simmental Assn 1 Simmental Way Bozeman, MT 59715
South Devon	North American South Devon Assn PO Box 68 Lynnville, IA 50153
Texas Longhorn	Texas Longhorn Breeders Assn of America 2315 N Main, Suite 402 Ft Worth, TX 76106

DAIRY

Ayrshire	Ayrshire Breeder's Assn 2 Union St Brandon, VT 05733
Brown Swiss	Brown Swiss Cattle Breeder's Assn Box 1038 Beloit, WI 53511
Guernsey	American Guernsey Cattle Club PO Box 27410 Columbus, OH 43227
Holstein-Fresian	Holstein-Fresian Assn of America 1 Holstein Place Brattleboro, VT 05301
Jersey	American Jersey Cattle Club PO Box 27310 Columbus, OH 43227
Milking Shorthorn	American Milking Shorthorn Society 1722JJ S Glenstone Ave Springfield, MO 65805

SHEEP

Cheviot	American Cheviot Sheep Society RR1 Box 100 Clarks Hill, IN 47930
Columbia	Columbia Sheep Breeders Assn Of America PO Box 272 Upper Sandusky, OH 43351
Corriedale	American Corriedale Assn Box 29C Seneca, IL 61360
Cotswold	American Cotswold Record Assn 282 Meaderboro Rd Rochester, NH 03867
Delaine Merino	American & Delaine Merino Record Assn 1193 Twp Rd 346 Nova, OH 44859
Dorset	Continental Dorset Club PO Box 577 Hudson, IA 50643
Finnsheep	Finnsheep Breeders Assn PO Box 34303 Indianapolis, IN 46234

Hampshire	American Hampshire Sheep Assn PO Box 345 Ashland, MO 65010
Karakul	American Karakul Registry RFD 1 Box 179 Rice, WA 99167
Lincoln	American Lincoln Assn 64546 Island Dr Deer Island, OR 97054
	National Lincoln Sheep Breeders Assn RR 6 Box 24 Decatur, IL 62521
Montadale	Montadale Sheep Breeders Assn PO Box 44300 Indianapolis, IN 46244
Natural Colored Wool	Natural Colored Wool Growers 18150 Wild Flower Dr Penn Valley, CA 95946
North Country Cheviot	American North Country Cheviot Sheep 833 Fall Creek Rd Longview, WA 98632
Oxford	American Oxford Down Record Assn Rt 4 Ottawa, IL 61350
Panama	American Panama Registry Assn Rt 2 Jerome, ID 83338
Polypay	American Polypay Sheep Assn Rt 2 Box 2172 Sidney, MT 59270
Rambouillet	American Rambouillet Sheep Breeders Assn 2709 Sherwood Way San Angelo, TX 76901
Romney	American Romney Breeders Assn 4375 NE Weslinn Dr Corvallis, OR 97333
Shropshire	American Shropshire Registry Assn PO Box 1970 Monticello, IL 61856
Southdown	American Southdown Breeder's Assn Rt 4 Box 14B Bellefonte, PA 16283

Suffolk	American Suffolk Sheep Society 1115 N Main Logan, UT 84321
	National Suffolk Sheep Assn PO Box 324 Columbia, MO 65201
Targhee	·U.S. Targhee Sheep Assn Rt 2 Box 6 Jordan, MT 59337
Tunis	National Tunis Sheep Registry RD 1 Wayland, NY 14572

SWINE

Berkshire	American Berkshire Assn PO Box 2436 1769 US 52 N West Lafayette, IN 47906
Chester-White	Chester White Swine Record Assn 1803 W Detweiller Dr Peoria, IL 61614
Duroc	United Duroc Swine Registry 1803 W Detweiller Dr Peoria, IL 61614
Hampshire	Hampshire Swine Registry 1111 Main St Peoria, IL 61606
Landrace	American Landrace Assn PO Box 2340 West Lafayette, IN 47906
Poland China	Poland China Record Assn PO Box 2537 West Lafayette, IN 47906
Spotted	National Spotted Swine Record PO Box 249 Bainbridge, IN 46105
Yorkshire	American Yorkshire Club Box 2417 West Lafayette, IN 47906

Resources

Cow-Calf Management Guide--Cattleman's Library...$37.00
available through your local county extension office or
  Bulletin Department
  Cooperative Extension
  Cooper Publications Building
  Washington State University
  Pullman, WA 99164-5912

The Sheepman's Production Handbook by Sheep Industry
Development Program, Inc.
200 Clayton St.
Denver, CO 80206

Sheep Housing and Equipment Handbook 3rd Ed., 1982
Midwest Plan Service
Ames, IA 50011

Index

ORDER FORM

Pine Forest Publishing
314 Pine Forest Road
Goldendale, WA 98620
Telephone (509) 773-4718

Please send _____copies of *LIVESTOCK SHOWMAN'S HANDBOOK* to:

Name _____

Address _____

State _____ Zip _____

Phone _____

Find enclosed a check or money order for _____, at $14.95 per book plus $1.50 postage and shipping ($16.45 each). Payment must accompany order.
Make checks payable to *Pine Forest Publishing*.
Washington residents: Please add sales tax ($1.15)
_____ Please send a free copy of group rates available for orders of 5 books or more.

✂ --

ORDER FORM

Pine Forest Publishing
314 Pine Forest Road
Goldendale, WA 98620
Telephone (509) 773-4718

Please send _____copies of *LIVESTOCK SHOWMAN'S HANDBOOK* to:

Name _____

Address _____

State _____ Zip _____

Phone _____

Find enclosed a check or money order for _____, at $14.95 per book plus $1.50 postage and shipping ($16.45 each). Payment must accompany order.
Make checks payable to *Pine Forest Publishing*.
Washington residents: Please add sales tax ($1.15)
_____ Please send a free copy of group rates available for orders of 5 books or more.